Moon Lander

Smithsonian History of Aviation and Spaceflight Series

Dominick A. Pisano and Allan A. Needell,
Series Editors

Since the Wright brothers' first flight, air and space technologies have been central in creating the modern world. Aviation and spaceflight have transformed our lives—our conceptions of time and distance, our daily routines, and the conduct of exploration, business, and war. The Smithsonian History of Aviation and Spaceflight Series publishes substantive works that further our understanding of these transformations in their social, cultural, political, and military contexts.

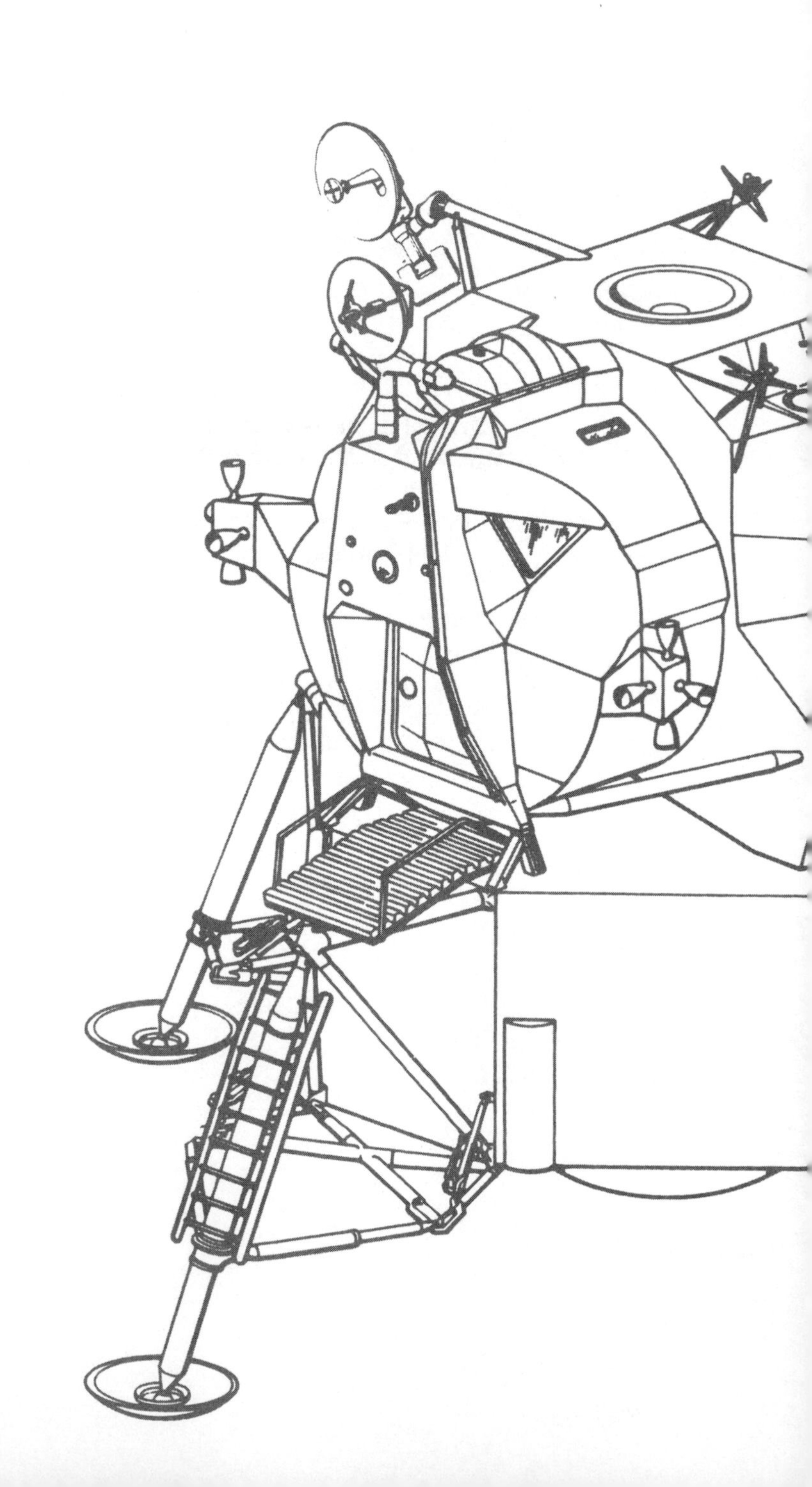

HOW WE DEVELOPED THE APOLLO LUNAR MODULE

THOMAS J. KELLY

SMITHSONIAN INSTITUTION PRESS
Washington and London

Copy editor: Karin Kaufman
Production editor: Robert A. Poarch
Designer: Chris Hotvedt

Library of Congress Cataloging-in-Publication Data
Kelly, Thomas J., 1929–
Moon lander : how we developed the Apollo lunar module / Thomas J. Kelly.
p. cm. — (Smithsonian history of aviation and spaceflight series)
Includes bibliographical references and index.
ISBN 1-56098-998-X (alk. paper)
1. Lunar excursion module. 2. Project Apollo (U.S.). I. Title. II. Series.
TL795.K45 2001
629.44—dc21 00-063728

British Library Cataloguing-in-Publication Data is available

Manufactured in the United States of America
06 05 04 03 02 5 4

∞ The paper used in this publication meets the minimum requirements of the American National Standard for Information Sciences—Permanence of Paper for Printed Library Materials ANSI Z39.48-1984.

To Joan, whose loving support
made my lunar adventure possible.

Contents

Illustrations

Acronyms

Acronym	Full Name	Definition
ACE	Automated checkout equipment	Computerized system for testing spacecraft
AEA	Abort electronics assembly	Computer used in AGS
AGS	Abort guidance system	Backup guidance and control in LM
AIAA	American Institute of Aeronautics	Aerospace engineering professional society and Astronautics
ALSEP	Apollo lunar surface experiments	Experiments deployed on the lunar surface package
AMPTF	Apollo Mission Planning Task	Prepared mission plans (time lines) and the Force design reference mission
APS	Ascent propulsion system	Ascent rocket engine and tanks
ASA	Abort sensor assembly	Inertial reference sensors for AGS
ASDTP	Apollo Spacecraft Development Test Plan	Overall program test plan
ASPO	Apollo Spacecraft Program Office	NASA-Houston program management group for the Apollo spacecraft
ATCA	Attitude and translation control	Flight-maneuver hand controllers assembly
BAFO	Best and final offer	Contractor's last offer in competition

CARR	Customer Acceptance Readiness Review	Formal review authorizing spacecraft delivery to NASA
CDR	Critical Design Review	Approval of detailed design
CM	Command module	Launch and reentry spacecraft
CRT	Cathode ray tube	Monitor for computer processed data
CSM	Command/service modules	CM and SM mated
DECA	Descent engine control assembly	Descent-engine controller
DEDA	Data entry and display assembly	AGS data entry keyboard
DFI	Development flight instrumentation	Added measurements for engineering data
DOD	Department of Defense	U.S. Department of Defense
DPS	Descent propulsion system	Descent rocket motor and tanks
DR	Discrepancy report	Chits written against problems during reviews and flights
DRM	Design reference mission	"Typical" mission plan and time line to establish design requirements
ECS	Environmental control system	Oxygen and thermal control
EMI	Electromagnetic interference	Unintended electrical or magnetic signal distortion
EO	Engineering order	Documentation authorizing drawing changes
EOR	Earth-orbit rendezvous	Mission mode with two Earth launches and rendezvous and assembly in Earth orbit
EPS	Electrical power system	Batteries and power distribution
ETRB	Executive and Technical Review Board	Corporate oversight board for LM
EVA	Extravehicular activity	Spacewalks outside the spacecraft
FITH	Fire in the hole	Igniting LM ascent engine atop the descent stage
FTA	Flammability test article	Boilerplate LM cabin for flammability tests

GNC	Guidance, navigation, and control	Guidance and attitude system control
ICD	Interface control document	Drawings/specifications defining interfaces between spacecraft
IFM	In-flight maintenance	Component replacement in flight
ISS	International Space Station	Large space station produced by NASA and an international team
KSC	Kennedy Space Center	Launch site at Cape Canaveral, Florida
L/D	Lift-to-drag ratio	Index of aerodynamic maneuvering capability
LEM	Lunar excursion module	NASA's early name for the lunar module
LES	Launch escape system	Escape rocket atop CM at launch
LGC	LM guidance computer	Computer for LM guidance and control
LM	Lunar module	Lunar landing spacecraft
LMP	LM mission programmer	Special programmer for controlling unmanned LM flight
LMS	LM mission simulator	Ground-based LM flight simulator
LOR	Lunar-orbit rendezvous	Mission mode in which LM rendezvous with CM in lunar orbit
LRV	Lunar roving vehicle	Electric-powered car used on later missions
LTA	LM test article	Full-scale, partially equipped LM for special tests
LTM	LM test model	Reduced-scale test model LM or components
MCC	Mission Control Center	Room in NASA-Houston where the flight director and controllers directed flight missions
MER	Mission Evaluation Room	Room in NASA Houston Building 45 where contractor engineers supported flights

MET	Modular equipment transporter	Hand-drawn cart for carrying equipment on lunar surface
MOCR	Mission Operations Control Room	Main control room in NASA-Houston MCC
MSC	Manned Spacecraft Center	NASA center at Houston responsible for developing the Apollo Spacecraft (now called Johnson Space Center [JSC])
MSFC	Marshall Spaceflight Center	NASA center at Huntsville, Alabama, responsible for booster-stage development
MSFN	Manned spaceflight network	Ground-based radar net work for tracking spacecraft in flight
MSR-B	Mission Support Room	Flight support room at Grumman-Bethpage –Bethpage
NAA	North American Aviation	Apollo spacecraft (CSM) contractor
NASA	National Aeronautics and Space Administration	Federal agency responsible for civilian spaceflight
O/F	Oxidizer/fuel ratio	Required ratio for rocket firing
O&C	Operations and Checkout	Building at KSC for space-craft launch preparation
OCP	Operational checkout procedure	Step-by-step test procedure
OMS	Orbital maneuvering system	Space shuttle's propulsion and RCS
PCM	Pulse code modulation	Digital sensor readout technique
PD	Preliminary Design	Engineering group for systems studies and proposals
PDR	Preliminary Design Review	Meeting for approval of preliminary design
RCS	Reaction control system	Small rockets that control attitude and maneuvers
RFP	Request for proposals	NASA's invitation to contractor bids
RTG	Radioisotope thermoelectric generator	Isotope power source for ALSEP

S/CAT	Spacecraft Assembly and Test	LM final assembly and test organization
S&C	Stabilization and control	Controls flight attitude and maneuvering
SCAPE	Self-contained air protection equipment	Propellant-resistant suit
SEB	Source Evaluation Board	Proposal evaluators
SIM	Scientific instrument module	Moon observation sensors aboard CSM
SLA	Spacecraft/LM adapter	Structure housing LM on Saturn
SM	Service module	Consumable and propul sion module for CM
SPAN	Spacecraft Analysis	Room in Mission Control from which spacecraft contractors supported flights
SPS	Service propulsion system	Rocket engine system in SM
SSA	Source selection authority	Senior source selection official
STG	Space Task Group	Organization at NASA-Langley studying manned spaceflight
STM	Spacecraft team manager	Responsible for a specific spacecraft (LMor CSM)
SWIP	Super Weight Improvement Program	Major weight reduction effort
TPP	Total package procurement	Fixed-price purchase of weapon system
TPS	Test preparation sheet	Detailed procedures for readying LM for tests and checkout
TSM	Technical staff meeting	Daily LM engineering meeting
VAB	Vehicle Assembly Building	Facility at KSC for assembling the Apollo booster and spacecraft "stack"
VHF	Very high frequency	2.5 to 3 megahertz for LM
VIP	Very important persons	Special viewing area in Mission Control

1

A Difficult Delivery

Bone weary, I shuffled across the Grumman parking lot after midnight under a full Moon, its light throwing my shadow ahead on the asphalt. I looked up, hesitated, stopped, and stared. The Moon looked close enough to touch, but it seemed impossibly far away. Almost a quarter of a million miles.

Galileo named the shadows on the Moon Maria because he thought they were oceans. The Egyptians plotted its phases. The Druid builders of Stonehenge calculated its rise and fall. Down through the ages, people have stared at the Moon's mysterious surface and wondered what was there. And now I—an obscure young engineer working for a small Long Island airplane company—I was going to provide the answers? It seemed absurd to me.

But the next morning I was back in the shop supervising a team preparing the lunar module spacecraft for its first test in space, a vital step in putting men on the Moon. Whenever I thought we had passed a crisis and the pressure of readying the craft would ease, we would find yet another broken wire in the spacecraft and have to stuff yet another two hours of splicing and retesting into our twenty-four-hour work schedule.

We were scrambling to complete the required tests, equipment installations, and inspections on LM-1, which was specially equipped to be flown unmanned in Earth orbit to verify the operation of the craft's major systems in space. It would be launched alone atop a Saturn 1B booster rocket, a smaller, two-stage precursor to the mighty three-stage Saturn 5, which would launch the combined Apollo spacecraft (command module [CM], service module [SM], and lunar module [LM; pronounced "lem"]) to the Moon.

The lunar module was my baby; I had led the technical studies and proposals that over a period of two and a half years resulted in Grumman's winning a NASA competition to design, develop, and build it. (After the first Moon landing, I was called "the Father of the LM.") It began in early 1960, when I was assigned to the Space Group in Grumman's Preliminary Design (PD) Department and told to find out about NASA's plans to put men on the Moon. We put company-funded effort into both studying the missions NASA was investigating and designing conceptual spacecraft to accomplish them. When President Kennedy announced in May 1961 that America would indeed put men on the Moon, Grumman had in place a knowledgeable technical study group that could prepare a competitive proposal.

Nevertheless, our company fathers decided that Project Apollo was too big for our small aircraft company. They directed that we work as subcontractor to a larger prime contractor, General Electric (GE). We gained a place on GE's Apollo proposal team as the command module crew compartment subcontractor. Joe Gavin headed Grumman's effort as vice president and program director, and I led the technical work as project engineer. As it turned out, the Apollo spacecraft competition was won by North American Aviation, but Grumman gained a great deal of knowledge and insight into the program as a result of the GE proposal effort.

We bounced back by helping NASA resolve how to perform the lunar landing and exploration mission, developing the rationale behind favoring the lunar-orbit rendezvous (LOR) approach. My group worked out many variations on both the mission and the design of the lunar module, the unique spacecraft that LOR required. When NASA decided upon LOR for Apollo, Grumman was well prepared. I led the technical proposal effort in the fall of 1962 that won the job for us over seven competitors.

After contract award, as project engineer and then engineering manager, I was the chief engineer for LM at Grumman. I directed the LM engineering team, which grew from an initial few dozen people to about three thousand at its peak, as LM progressed from preliminary to detailed design and, finally, to manufacturing assembly and testing. In February 1967, as the critical activities shifted from engineering to assembly and testing, I was put in charge of LM Spacecraft Assembly and Test (S/CAT).

As S/CAT transformed the LM from lines on paper into functional hardware, I strengthened the dedicated but undermanned crew by adding engineers and providing additional training and supervision. The work was slow and frustrating because we were developing many new things at once: the LM spacecraft itself, the critical manufacturing processes needed to build it, the ground-support equipment and acceptance-checkout equipment that tested it, and the detailed written procedures that controlled and documented every action we performed on it.

We were under pressure to solve problems quickly and to complete the preparations for LM-1's delivery to the launch site at the Kennedy Space Center (KSC) in Florida. The Apollo Project's goal, set by President Kennedy, was to land men on the Moon before the end of the decade, less than three years away. Although we still could not keep to promised schedules, we had completed most of the tests and installations required for LM-1's delivery. The LM-1 Customer Acceptance Readiness Review (CARR) was scheduled for 21 June 1967 at Grumman's main facility in Bethpage, Long Island, New York.

The CARR was a formal review lasting one week, attended by more than two hundred NASA engineers and their Grumman counterparts. Its purpose was to establish LM-1's compliance with all NASA and Grumman requirements by reviewing the documentation accumulated on the spacecraft as it was built and tested in S/CAT and at our suppliers. The LM-1 itself would be physically inspected by the review team.

The review team was organized into panels according to technical disciplines and systems. As the panels waded through the assembled documentation, they generated discrepancy reports, called DRs or "chits," for any apparent deviation from requirements, which were reviewed each night by the panel co-chairmen (NASA and Grumman) and "dispositioned" by categories.[1] Only chits in the "Unresolved" category were presented to the CARR Board for decision on the final day of the review.

The CARR was semiorganized chaos. Finding adequate work space and meeting rooms for more than four hundred members of forty panels was a challenge, as was the effort required to gather, reproduce, and index the thousands of documents that had to be scrutinized. A review center was established in three large conference rooms on the first floor of Plant 25, the primary LM engineering building. Many of the reviewers physically inspected items on LM-1, which remained in the Assembly and Test clean room for final outfitting. I made sure that S/CAT completed all remaining LM-1 items required before delivery.

Grumman's senior vice president, George Titterton, handled the meeting arrangements. A five-foot-two blustery man who was determined to prove he could take on any challenge, he decided to pattern the CARR on the annual Grumman stockholders' meeting. The day before the meeting, airplanes were moved out of the Plant 4 flight operations hangar onto the adjacent apron and a wooden dais was assembled at one end of the cavernous space. The hangar had a vaulted elliptical roof, and its end walls nested ceiling-high sliding doors that could be fully opened to allow the crowd to assemble and disperse easily. Several hundred folding wooden chairs were set up in rows facing the dais, and a portable sound system and several large trumpet-shaped loudspeakers on collapsible metal stands were provided to penetrate the acoustically dead environment. Behind the last row of folding chairs there was a food catering area, complete with serving counters and dining tables and chairs. A large projection screen was positioned catty-cornered on the stage, where it could be viewed by both the CARR Board and the audience, and lecterns and microphones were provided for the board chairman and the briefer.

The CARR Board was chaired by Bob Gilruth, director of NASA's Manned Spacecraft Center (MSC) in Houston. NASA members of the board included

George Low, Apollo spacecraft program manager at Houston; Gen. Samuel C. Phillips, Apollo program director; Eberhard Rees, deputy director of the Marshall Spaceflight Center at Huntsville; Kurt Debus, director of Kennedy Space Center; and Chris Kraft, flight operations director at MSC. Grumman members included George Titterton; Joe Gavin, vice president of Space Programs; Ralph Tripp, LM program director; Bill Rathke, LM program manager; Edward Z. Gray, assistant to Grumman's president; and me. The senior NASA people flew into the Grumman airport at Bethpage in the morning.

We enjoyed a pleasant though crowded lunch, served on real china and flatware at the tables in the catering area. Shortly before one o'clock we took our places around the large U-shaped table on the stage. As we mounted the stage we were hit by a solid roar of sound rising from below. Above the drone of many conversations rose the clatter of china and clinking of table silver. Three times I tried to talk to Chris Kraft across the table but gave up, hoping that the din would subside when the microphones were activated.

Gilruth called the meeting to order. Normally soft-spoken, he had to shout into the microphone before the hubbub subsided. Then the sound system broke into an ear-piercing feedback squeal. As the participants filed into the rows of seats the babble of conversation diminished, but on the stage the clanging of the tableware seemed louder and more distracting. Titterton motioned for a young engineer to quiet the caterers.

We began the lengthy agenda. The list of "Unresolved" items to be presented to the board had been winnowed to about forty. A brief of each item was presented by the person who wrote the discrepancy report, then a senior member of the DR Review Board commented on the brief. The CARR Board discussed the item and Gilruth solicited opinions around the table. Although Gilruth rendered the final verdict, he sought to obtain consensus with Gavin.

This logical but arduous process was hampered by the noise level and the hangar's sound-swallowing acoustics. We could hardly hear one another across the table, even when we shouted into the microphone. Another loudspeaker was added on the front of the stage, not facing the audience like the others, but turned around to face the board table, but it did not solve the problem.

Despite these difficulties the board soldiered onward. I was torn between suppressed glee at Titterton's growing discomfort and concern that at some point the NASA board members might simply get up and leave. Gilruth, Low, and Phillips were patient, repeating questions two or three times. Finally, though, Gilruth had had enough. He shouted into the microphone sarcastically, "If it's not asking too much, would you people in the audience please stay in your seats and be quiet, so the board can hear the speaker and each other?"

Gilruth's patience, good judgment, and common sense extended to the substance of the meeting also. As each unresolved DR was reviewed, he carefully considered the views of the concerned writer of the DR (the plaintiff) and of those who felt that the discrepancy could be explained away or was not really a problem (the defendants). He was quick to sense phony "cover your ass" DRs and disapprove them.

"We'd have to open up a sealed reaction control system to verify that?" he asked after one presentation, peering quizzically over his glasses.

The plaintiff nodded. "Yes, sir."

"And it's already installed and brazed into place?"

"Yes, sir."

"Then I say the cure is worse than the disease—if there *is* a disease. Next item, please."

Whenever the problem in question seemed real, all of us on the board strove for a practical solution that would not unduly disrupt the spacecraft systems or the schedule. Some equipment could readily be removed and retested or replaced before shipment from Bethpage. Other equipment was planned for retest or replacement after delivery to KSC. In some cases we simply compromised on questionable items. We decided, for example, that one troublesome sensor, which had failed and been replaced three times, was not mandatory for flight. "This ascent engine oxidizer injector inlet pressure measurement is intended to provide data for comparison with ground tests if we encounter combustion instability in flight. It's not required for operating the spacecraft," I explained during the board's discussion after a DR presentation.

"I think it's a highly desirable measurement, but I wouldn't call it mandatory. It's too bad you couldn't make it work, but you've tried three times, and I think that's enough," was Gilruth's judgment.[2] In no case did the board consciously take a risk that could affect the success of the mission.

The board had displayed remarkable patience and adaptability and was making steady progress down the agenda when an overwhelming, deep-throated roar engulfed the hangar. Titterton jumped up, angry and embarrassed. A flight line engine test crew had started testing a new Gulfstream II executive jet outside the hangar, the usual place for such tests.

Using hand signals and shouting into the microphone, Gilruth declared a break in the meeting. Titterton first ordered the four ceiling-high hangar doors closed. That hardly affected the noise level, but the temperature soared. As Low and Gilruth started toward him, probably to ask that the meeting be adjourned, Titterton hurried outside to order the Gulfstream captain to halt his tests. I watched the NASA officials desultorily thumbing through some of the briefing handout material and looking at their watches every minute or two. Surely in another few minutes they would leave, the CARR Board would have to find a new date in their busy calendars, and we would lose a week or more of precious LM schedule.

Suddenly the din ceased. Silence was like a release from pain. Titterton strode briskly in from the runway apron, sweating, his face a rosy pink, his glasses fogging. The hangar doors were reopened, and air circulated. A cathedral silence settled over the crowded hall as we held our breath. Gilruth leaned into the microphone and, from newly formed habit, shouted in his reedy voice, "I guess that's over." Then, realizing he was shouting into silence, he dropped into a normal tone to finish: "George assures me it won't happen again."

Four hundred pairs of eyes turned to Titterton, who excelled in having the last word. This time he could only muster a wan smile and nod assent.

The CARR Board resumed its deliberations. Our working conditions steadily improved as the DRs for each LM-1 system were dispositioned and the panel members and others concerned with those items left. The crowd thinned, the background noise lessened, and the temperature dropped. After several hours of watching the board, the panel members gained a sense of the probable outcomes of the DRs. Plaintiffs and defendants began to negotiate "out of court" settlements. Our pace accelerated, and the impossible agenda was completed shortly after 8:00 P.M.

Then came the grand moment: Gilruth declared that the CARR Board had found LM-1 ready for formal acceptance and delivery to KSC, subject to the satisfactory completion of the assigned action items. (Before and after delivery Grumman was still responsible for LM-1.) My notebook entry for that day, after several pages covering the agreements for resolving DRs, concludes triumphantly in large block letters: OKAY TO SHIP.[3]

Grumman and NASA people worked all night and into the next day completing action items required before shipment and updating inspection records and delivery papers. The last form to be signed was the DD-250, the document by which the government accepts the product from the contractor and takes ownership. There were smiles all around when John Johanson, NASA manager at Bethpage, signed on behalf of the space agency.

In the afternoon the LM-1's ascent and descent stages were carefully secured within their specially designed shipping containers and loaded inside the oversized belly of a modified Stratocruiser NASA used to transport spacecraft for the Apollo Project. With pride and a great sense of relief, I watched the huge aircraft make a lumbering takeoff, using most of the mile-long Bethpage runway. The first LM was on its way to fly in space!

We had come a long way since first learning of NASA's early studies in 1960 of the possibility of landing men on the Moon. Ahead lay the unmanned LM-1 flight and the manned flights in Earth orbit and lunar orbit preceding the first manned lunar landing by Neil Armstrong and Buzz Aldrin in Apollo 11. Even as we delivered LM-1, my colleagues and I were preparing several other LMs in the Assembly and Test area, including LM-5, which would become Armstrong's trusty lander Eagle. I certainly had not foreseen the massive effort involved and the relentless striving for perfection, which was our way of assuring that we had done our very best to make Grumman's part of the mission a success. It had been an exhilarating and totally absorbing effort. The outcome was still uncertain, but those first LM stages winging south were a major milestone, promising more achievements to come.

1

Winning

2

We Could Go to the Moon

Grumman's Plant 5 in Bethpage was a red-brick building so long that it seemed less than its three stories, and the narrow stripe of windows on the third floor only accentuated its horizontal appearance. It was separated from the parking lot by a neat strip of putting-green grass. The U.S. Navy built Plant 5 for Grumman during World War II. At that time it housed the experimental airplane shops and the engineers busy devising more powerful and deadly versions of the Wildcat and Hellcat fighters and the Avenger torpedo bomber, the navy's great workhorses in the Pacific air war against Japan. Since then it had been modified and expanded many times until it was a labyrinth of corridors, alcoves, hidden shops, and laboratories. Visitors and new hires often became lost in it, but to the initiated like me, its weblike convolutions added to the sense of belonging to a secret society. I found it a pleasant place to work.

Seven Years Earlier

I worked in an inner sanctum, the Preliminary Design mezzanine, also called the "Hanging Gardens," a strange architectural afterthought of cinder-block suspended above the workbenches and drill presses of the experimental shops. Within this fluorescent-lighted, windowless compound, time stood still and total concentration was possible.

I had joined Grumman's Propulsion Section in 1951 after graduating from Cornell in mechanical engineering. For five years I analyzed and designed inlet and exhaust systems for the supersonic ramjet-powered Rigel missile and for Grumman's jet aircraft, including the advanced F11F-1F supersonic fighter. In 1956 I was called into active duty with the U.S. Air Force, at the Aircraft Laboratory at Wright Air Development Center, where I worked on the jet propulsion systems for the air force's many new aircraft under development, including the F104 Starfire, the F106, and the B-58 Hustler supersonic bomber.

During my air force tour of duty, the Russians launched Sputnik, ushering in the space age. I became fascinated with the idea of engineering vehicles for space, and as the end of my air force time drew near I interviewed aerospace companies for opportunities in space engineering. In 1958 I joined Lockheed in Sunnyvale, California, to work on space propulsion systems and rockets. My former colleagues lured me back to Grumman a year later with the promise that they were about to mount a major effort to get into space technology.

To investigate opportunities in space, Grumman formed the Space Steering Group in the Preliminary Design Department under Al Munier. A thin man of medium height with a ready smile but quick temper, Munier delighted in challenging his young engineering charges to think big and do the impossible. The group was expanded to about half a dozen engineers and made part of the Advanced Space Systems Department of PD when Grumman found that numerous space-system development opportunities were unfolding within NASA. Grumman management was cautious about venturing into the relative unknown of space systems, but there was much promising business and our competitors' credentials did not seem any better than ours. Gradually the Grumman drive for space-system development became more aggressive.

For two hectic months I had been the Propulsion Section's PD representative on the proposal to NASA for the orbiting astronomical observatory (OAO), a large telescope in Earth orbit, the forerunner of the later, even larger Hubble Space Telescope. The OAO operated successfully for more than eight years, yielding much new astronomical data. The Propulsion Section's contribution covered the reaction control system, which controlled the satellite's attitude and position in space using small rocket thrusters, and the thermal control system, which maintained stable internal temperatures by means of insulation and surface finishes. The OAO's mission required pointing and stabilizing the spacecraft to 0.1 seconds of arc, equivalent to sighting telescopic cross-hairs on the Washington Monument from the Empire State Building. Grumman had demonstrated such accuracy in the laboratory. If successful in space, it would be a major technical achievement. To support its proposal, Grumman committed to a major investment in space facilities, including a high-bay, clean-room spacecraft assembly area and a large space-simulating thermal-vacuum chamber.

After the proposal was submitted I returned to engineering activities in Propulsion, thinking I might be offered a job on OAO if we won, even though I had not been named to any of the key positions in the proposal. One day Al Munier brought me into his office and told me the company did not want me tied up with OAO. They felt I would be more valuable exploring other space opportunities.

I was disappointed because I had become fascinated with the technical challenges posed by the OAO, but I was also flattered that Munier considered me a key person in Grumman's space future and was trying to use me effectively. A few days later Munier again summoned me into his office.

"Well, Tommy," he teased, "I think we've found just the job for you."

"What is it?"

"How would you like to design a spaceship to take men to the Moon?"

"Are you kidding? Look, Al, I didn't come back to Grumman from Lockheed to chase pipedreams. You guys said you were serious about getting into space."

Munier's face darkened. I had overstepped the unspoken bounds with my flippant response.

"Well, *Mister* Kelly," he snapped, "that just shows how little you know about the space business. NASA is planning a major project to send men to the Moon. It's called Apollo, and they're going to spend billions of dollars on it. I don't see why Grumman couldn't be a part of it, if I can get certain pigheaded engineers to work with me."

Munier said he would introduce me to Tom Sanial, who had been studying NASA's plans for Apollo. He wanted me to work with Sanial to figure out what Grumman's part should be in the Moon program. I left Grumman that night in a state of anticipation. Was it hope or intuition? The old tune "This Could Be the Start of Something Big" was ringing in my head.

Tom Sanial was destined for big things at Grumman. Although only twenty-seven, he had been tapped for the elite Preliminary Design Group as structural design engineer for the Mercury proposal. He became the assistant project engineer, due to his ability to visualize the entire design and integrate it into engineering drawings and three-dimensional illustrations as well as the brilliant design work he had done for Joe Gavin as a structural designer on the F9F Panther and Cougar.

Grumman's Mercury proposal to NASA won the eleven-company competition, a nose ahead of the runner-up, McDonnell Aircraft. The elegantly simple design—the blunt-faced, one-man reentry capsule protected from aerodynamic heating by a base heat shield of thermoplastic resin material and side shingles of beryllium—was confirmation of the ingenuity of Grumman engineering. None of the competitors had ever tried to design a manned space capsule before, and the youthful Grumman team won this contest of imagination and practical design.

Before final selection of its contractor, however, NASA consulted with the U.S. Navy, the primary customer of the two top-ranked competitors. The navy said that Grumman was heavily loaded with work on recently awarded contracts for the A6 Intruder attack bomber and the E2 Hawkeye early warning aircraft. Influenced by this advice, NASA awarded the Project Mercury contract to McDonnell.[1]

After the loss, Sanial expected to join one of the new aircraft projects and was exploring the possibilities with his friends in A6 engineering. He was surprised when Al Munier urged him to remain in Preliminary Design as a member of the newly formed Space Sciences Group.

"Space is where the future is, Tom," Munier told him. "We're going to keep trying until Grumman is really in the space business. It's a great opportunity for a young fellow like you."

Munier was persuasive, and Sanial was fascinated with the idea of designing spacecraft. Everything about it was new: the environment, the missions, the systems requirements. A young engineer with imagination and talent had as good a chance of success as a seasoned aircraft designer.

After four months chasing an elusive will-o'-the-wisp called Apollo through the halls of NASA, Sanial was not sure he had made the right choice. Project Apollo was gaining momentum, he was sure of that, but except for Al Munier, who seemed, he thought, to go overboard on any new and different idea, no one else in the company appeared to pay much attention. He had begun to lose his own sense of urgency, working normal hours and sometimes taking a day off to take his wife and two young daughters to the beach or for a sail. Maybe he should relax and enjoy life more, instead of always striving to achieve bigger and better things. But then Al Munier brought him an unexpected new partner.

I followed Munier into Sanial's small cubicle deep inside the Preliminary Design mezzanine. We were greeted by a tall, trim fellow with a long, freckled face. His hair was more gray than black, a contrast to his boyish, youthful face. He was neatly dressed in a blue blazer jacket and tie, even hidden away in this obscure den.

"Tom Sanial, meet Tom Kelly," Munier said breezily. "He's here to work with you on Apollo."

Munier left us alone together in awkward silence as we sized each other up. Were we to be partners or rivals? Sanial's innately generous nature took over. Soon he was showing me his files and describing what he had learned of the Apollo Project. He was persuasive and persistent as he related the steady growth of the program shown by NASA's statements, budgets, and planning documents. He had become convinced that NASA was seriously planning a manned exploration of the Moon.

I moved into the desk next to his in the cinder-block cubicle, which had six desks in all, and pored over the documents he had collected, along with his technical notes and trip reports. In the spring of 1960 we visited NASA Headquarters and the Langley and Lewis Research Centers, where I met DeMarquis "Dee" Wyatt at Headquarters and George Low at Lewis. They spoke in guarded terms, making clear that the manned lunar mission was in an early phase of internal review and was not part of NASA's firm plans. Their intense interest, however, indicated that the program was a serious possibility. I concluded there was at least an even chance that Apollo might happen, and that Grumman had better get prepared.

Suddenly our company was a full-fledged player in space: Grumman won the OAO competition. Eight aerospace companies had fought for this major NASA program, with its possibility of discovering new worlds in astronomy, and Grumman, known mainly for its World War II Wildcats and Avengers, had been a long shot. At the OAO victory party, some of my colleagues asked if I would be joining them on the project.

"I don't think so," I replied. "Gavin and Munier want me to work on Apollo."

"What's that?" they chorused.

"It's a program of manned exploration of the Moon."

"Send men to the Moon? Are you crazy?" one engineer scoffed.

"NASA seems serious about it. They're planning to spend billions of dollars to make it happen."

"Come on, Tom, that's Buck Rogers stuff. Towl and Titterton aren't going to stick Grumman's neck out for something like that. Face it—it's amazing that we were even allowed to bid on OAO. We wouldn't know where to begin a trip to the Moon." When he walked away shaking his head and laughing, I wondered if my career was headed for oblivion.

Now that Grumman was a serious space contender, Apollo became a more urgent business objective. NASA convened an Apollo industry planning session in late July 1960 that was attended by more than thirteen hundred people. At this conference they announced their interest in a manned lunar landing and showed the results of their engineering studies. Although not yet an official program, NASA interest in Apollo was very serious.

Sanial and I convinced Munier that we should form a small study group to educate ourselves about manned lunar missions. Munier spoke to Walter Scott, chief of Preliminary Design, who approved budget for ten people. We recruited the specialists we needed in aerodynamics, orbital mechanics, flight dynamics, and control and weights. Sanial and I were co-leaders of the study. He would also handle structural design while I covered propulsion.

We read everything published about Apollo and lunar missions in general and formulated a plan for a manned lunar mission feasibility study. In October 1960 we submitted this study plan as a proposal to NASA in an industry competition for the launch and lunar trajectory phases of the mission. (The overall plan for accomplishing the mission was not yet addressed.) We lost to Convair, General Electric, and Martin, but Grumman management (Walter Scott and Joe Gavin) decided to conduct our study anyway, using company money.

Two pressing issues for the study were the shape and aerodynamic characteristics of the manned reentry vehicle. Reentry through Earth's atmosphere from escape velocity (twenty-six thousand miles an hour) was much more difficult than from Earth orbital velocity, in terms of both the heat the vehicle must withstand and the guidance, navigation, and control (GNC) accuracy needed. The returning spacecraft had to enter a narrow window of allowable reentry angles, between 5.5 and 7 degrees, when it encountered the upper edge of Earth's atmosphere. Too steep an angle would cause the spacecraft to burn up; too shallow an angle would skip the craft off the top of the atmosphere to wander aimlessly through the solar system for eternity.

Even after reentry, when the spacecraft had decelerated to high supersonic speed in the lower reaches of the atmosphere, it must still maneuver to reach the landing area. A blunt body, like the Project Mercury capsule designed by Max Faget, a highly creative aeronautical engineer working for Bob Gilruth in the Space Task Group (STG) of NASA's Langley Research Center, had no aerodynamic maneuvering capability (lift-to-drag ratio [L/D] = 0). However, such a body spread the atmospheric frictional heating over the large area of the blunt end, making insulation and heat absorption easier. Lifting bodies, like the series of aerodynamic shapes developed by Alfred Eggers at NASA's Ames Research Center, had significant aerodynamic maneuvering capability (L/D = 0.5 to 3 or 4), which allowed them to glide several hundred miles to reach a specified landing area. But these shapes concentrated the aerodynamic heating on the vehicle's nose and the leading edges of its stub wings and tail, creating temperatures of several thousand degrees. Still, the lifting body made the job of the GNC system somewhat easier, because it could maneuver to compensate for errors in targeting the landing area from space. There were many interesting engineering tradeoffs and design compromises to be considered.

Our lifting body expert was Bob Lecat, an irascible and volatile pipe-smoking man of French heritage. A brilliant aerodynamicist, Lecat was devoted to his work, but at times that devotion caused him to appear unconcerned with the feelings of others. His gruff voice could often be heard lamenting the stupidity of Grumman management and of most of his engineering colleagues. With me he adopted an air of patient condescension, as though explaining Aerodynamics 101 to the village idiot.

Lecat designed a family of lifting-body shapes for analysis and comparison with the Mercury blunt body. His results were a principal element of our study report. In addition to analyses and wind-tunnel tests, Lecat built a model from balsa wood and tissue paper of his favorite design and flew it around the parking lot, getting derisive comments from aircraft engineers about its bloated appearance.

On 15 May 1961 we submitted our summary report to NASA, maintaining that a manned lunar mission was feasible. We presented the results of our reentry vehicle studies but left the choice of body shape open. Blunt bodies, lifting bodies, and winged vehicles all looked workable from a GNC error tolerance viewpoint, but the tolerances opened up with increasing L/D. Material and design difficulties increased with L/D, but the lifting bodies appeared feasible with advanced materials and manufacturing techniques.

The NASA-funded studies were much more thorough and elaborate than ours. Each of the three contractors spent more than a million dollars, although NASA had paid each only $250,000. At their midstudy reviews, each contractor chose a specific lifting body reentry vehicle. None recommended a blunt body, which annoyed Faget, who said in his Louisiana drawl that they just did not understand the problem. In their final reports, however, the contractors had gotten Faget's message: GE and Martin recommended blunt-body command modules, although Martin added flaps to give it some maneuverability. Martin's nine thousand–page report was the largest, the result of three hundred people working for six months and spending $3 million.[2]

We presented our results to NASA's Space Task Group a few days after the funded contractors' briefings. Considering the size of our investment—we averaged fifteen to twenty people on the study and probably spent less than $250,000—NASA gave us an attentive hearing and seemed interested in our results. Faget lectured us briefly on the advantages of blunt-body reentry vehicles, contending that they made reentry heating tolerable by known, practical materials. He thought we should drop lifting bodies from further consideration because they required state-of-the-art advances in high-temperature materials. But Faget also said that our study results were in line with NASA's own work. I felt that our effort on the study had been a good investment.

Throughout the LM studies and proposal I relied heavily on my deputy, Erick Stern, a brilliant analyst with a flair for debating and selling his ideas. Stern's work on the inertial guidance system for the Atlas ballistic missile at American Bosch Arma Corporation gave him practical experience with a critical system for spaceflight: guidance, navigation, and control. He broadened this to embrace the rapidly developing field of systems engineering pioneered by the air force and navy ballistic missile programs. I used him as a sounding board for all technical problems, whether or not they fit his background, and always received carefully reasoned answers.

Stern was tall and heavy-set, with a high forehead, thinning blondish brown hair, a pale face, and blue eyes behind a pair of thick glasses. He looked every inch an intellectual. He had grown up in Vienna and been brought to America by his parents before the war, and his trace of Viennese accent enhanced his professorial image. He worked his way through City College, where he earned a bachelor's degree in electrical engineering, and then, while working at Arma, attended New York University at night and obtained a master's degree.

Stern and I worked well together; we respected each other and both enjoyed serious technical discussions. Not everyone could keep up with Stern's racing mind, though, and he could lose patience with those who did not follow his arguments, no matter how high their rank. Some people took offense at Erick's frank manner, but it never bothered me, and only later did I realize that influential NASA people resented him. I felt that he was always trying to instruct and persuade, not put down others. His arguments had none of the dismissiveness I often sensed from Lecat. Not that Stern ever hesitated to point out where he thought I was wrong, even in the presence of others, but he was loyal to me, and he was skillful and dedicated to his work on the program, which overshadowed any minor carping about other facets of his personality.

In the meantime, a major debate was raging within the Kennedy administration to identify a space mission that would allow the United States to recapture the lead in space technology from the Soviet Union. The cold war had flared up, and the global competition between two diametrically opposed social systems broadened into the arena of outer space.

On 12 April 1961 the Soviets launched Yuri Gagarin, the first human to orbit Earth. A few days later the U.S.-sponsored Bay of Pigs invasion of Cuba failed ignominiously. President Kennedy was looking for a way to regain the initiative from the Soviets. He turned to his vice president, Lyndon B. Johnson, and the newly appointed NASA administrator, James E. Webb, for recommendations for a "space spectacular" that would clearly establish American preeminence.[3] The feasibility studies that we and the other contractors submitted fell on fertile ground.

Ten days after our report went into NASA, Kennedy declared before a joint session of Congress, "I believe that this nation should commit itself to achieving the goal, before this decade is out, of landing a man on the Moon and returning him safely to Earth. No single space project in this period will be more impressive to mankind, or more important for the long range exploration of space, and none will be so difficult or expensive to accomplish."

The dreaming and studying were over. America was going to the Moon! In our dreary Preliminary Design cubicle, Sanial and I could scarcely believe our good fortune. We grinned foolishly at each other: the roulette wheel had stopped on our number. What would it be like to build spaceships to fly to the Moon?

"I'll bet the real Apollo won't look like any of the vehicles we've studied," ventured Sanial.

"Why do you say that? Don't you think we've done a good job?" I challenged.

"Our study was okay as far as it went, but I'm sure we've just probed the obvious. There's still so much we don't know about how to fly to the Moon."

I had to agree with that. "You're right. We don't even know yet what we don't know."

The opportunity was there, now we just had to find a way to get a piece of it for Grumman and ourselves. If we succeeded, we would be part of history.

Losing Our Nerve

After President Kennedy's announcement, the scale and pace of Grumman's Apollo activities surged. Joe Gavin was put in charge of the Apollo studies as chief missile and space engineer. Sanial and I led the technical effort for the studies, reporting to Gavin and Munier. Our study group expanded from thirty to sixty people, including senior engineers such as Erick Stern and Bob Watson, a highly regarded GNC designer from the Dynamic Analysis Section in Engineering.

Gavin and Munier explored alliances with other companies to strengthen our lunar mission study effort. We were interested in both the technical expertise that specialists in technical systems could add to our studies and the contribution of skilled people and technology that other companies could make. Honeywell joined us first, in GNC, and Space Technology Laboratories (STL) joined next, in electronic systems and program management. Each brought new ideas and talent. We housed them, together with the added Grumman members of our expanded study group, in a large open area in the center of the PD mezzanine that had previously been reserved for major aircraft proposals. Munier, Sanial, and I and some of the other Grummanites stayed in the secluded Space Sciences corner adjacent to the main mezzanine floor.

The Apollo spacecraft request for proposals (RFP) was released to industry in July 1961, and in August NASA hosted an Apollo bidders' conference in Washington, D.C. In between the formal sessions, and at breakfast, lunch, and in the evening, contractors engaged in frantic mating rituals: exploring potential alliances for bidding on the Apollo program. Hasty assignations between contractor representatives took on a comic-opera air, with intense whispering behind the potted palms.

Joe Gavin, Saul Ferdman from Space Marketing, Bob Watson, and I all performed mating dances with potential teammates, either together or separately. When we compared notes at the end of the day, we sometimes found that a suitor was in bed with more than one company. We learned that there were at least as many companies who wanted Grumman to be a subcontractor as were willing to consider joining a Grumman-led team. Nobody got married, but there was a lot of heavy dating.

At the formal sessions, NASA described the scope of the Apollo Project and summarized the primary requirements that would be in the RFP. They planned to release the RFP within two or three weeks, allowing sixty days for response. The program was even bigger than we had envisioned. It included a three-man command module blunt-body reentry vehicle, a supporting unmanned service module containing consumables (such as propellants, helium, water, and oxygen), electric power, electronics and antennas, and a large restartable liquid-propellant rocket system. A launch escape rocket for the command module and a large cylindrical structural shell to support the service module atop the upper third stage of the Saturn booster rocket would also be supplied by the Apollo spacecraft prime contractor. Also required were the ground-support equipment (GSE) necessary to test and service the spacecraft and the integration of all the space and ground components into a functional system that could send men to the Moon and return them safely.

Apollo would be the biggest engineering job in history. Joe Gavin thought it was bigger than building the pyramids or inventing the airplane and would take every ounce of ingenuity for us to pull it off.

When Gavin reported to Grumman's top management that the Apollo RFP was imminent, he was asked to recommend what the company should do. Sanial and I prepared the technical briefing based upon our study results. We argued that Apollo was feasible and that we knew how to do it. We rehearsed our briefing several times before Gavin and Munier. Gavin and Ferdman gave the management briefing and budget request. They strongly recommended that Grumman bid as prime contractor. They estimated the budget necessary to prepare a winning proposal along with the facility and capital equipment investment required to perform the job: all huge numbers by Grumman standards.

The briefing was held in Grumman's mahogany-paneled board room to an audience seated around a large boat-shaped table. Grumman's president, E. Clinton "Clint" Towl (pronounced "Toll"), and executive vice president, William T. "Bill" Schwendler, were there. Both were founders, in December 1929, of Grumman Aircraft Engineering Corporation in Baldwin, Long Island. Clint Towl led the fledgling company in business management and administration. A reserved man of carefully chosen words and dry wit, he had curly brown hair and an intense stare behind small, gold-rimmed eyeglasses.

Bill Schwendler had been hand-picked by Leroy "Roy" Grumman at the predecessor Loening Corporation to be his engineering protégé and problem solver. Schwendler absorbed everything then known about aircraft design from his talented pilot-engineer-inventor mentor and became the company's chief engineer as Roy Grumman inevitably became more enmeshed in sales, customer relations, and general management. Bill recruited Vice President of Engineering Dick Hutton and other Grumman engineering leaders and guided the rapid growth of the Engineering Department. He had a ruddy, angular face, blond hair so thin and fine that he appeared almost bald, and vivid blue eyes, which he fixed unblinkingly upon his subject.

Two of Grumman's earliest employees were also there: Vice President of Engineering Richard "Dick" Hutton, who was chosen by Schwendler to succeed him as chief engineer, and Senior Vice President George Titterton, who directed Contracts, Business Administration, and Marketing. Chief Technical Engineer Ira Grant Hedrick, Grumman's most respected analytical engineer, rounded out the decision-making group.

On my first visit to the board room, I was awed by this corporate horsepower, some of whom I met for the first time. My briefing went smoothly as rehearsed. They put me at ease and asked reasonable questions to which, for the most part, I had answers: How fast is the spacecraft going when it reenters the atmosphere? What temperatures does this produce? Are there problems with meteorites and radiation in space? How will the crew know what maneuvers they must perform to reach the Moon?

Schwendler, Hutton, and Hedrick were intent on the technical briefing and wanted to thoroughly understand the basic spacecraft and mission design concepts. The idea of zero gravity caught their imagination; they likened it to a prolonged inside loop in an airplane. (Such a loop was later adopted for training astronauts.) "You'd have to secure everything so it wouldn't float around. I've seen clipboards, maps, and pencils wandering all over the cockpit during a loop that cancels out gravity," commented Dick Hutton.

Towl and Titterton bore in hard on the management briefing, worried about the size of the investment required by Grumman. After more than two hours of intense discussion, Sanial, Ferdman, and I were dismissed. Later that afternoon Al Munier returned to his office and called Sanial and me inside. The deliberations were still in progress, but he had heard enough before being asked to leave to suspect that a "no prime bid" decision was possible. He wanted to prepare us for that outcome.

"Al, they can't do that," I protested. "Not after all the good work we've done. We know more about how to do the Apollo mission and design the spacecraft than anyone in the county."

"For Pete's sake," Sanial chimed in, "some of our ideas are in the RFP. We're in the catbird's seat."

"You kids don't realize—you're asking them to bet their company." He fluttered his hands in a far-out gesture. "And on a project that sounds like science fiction. Most people in the company think we're all crazy."

"It's not impossible—you know that, Al. It's just something new, like supersonic flight. It's all just engineering and physics and know-how. Grumman has the talent and the tools. We could go to the Moon," I shouted.

"Take it easy, Tommy," Munier said in a patronizing tone. "Don't yell at me; I'm just the messenger. I just wanted to warn you what might happen upstairs." I went home burning with suppressed rage and hardly slept that night.

The next morning Joe Gavin assembled the team in the Space Sciences enclave and announced that Grumman would not bid Apollo as prime contractor. "Our senior management thinks its too big a job for us," Gavin said. "We'd be risking the whole company, and the jobs of everyone at Grumman, on this single project. It's not just the money involved. If the company failed before the world in this project, it would never recover. We'll have to find a berth on someone else's team."

"What do you think, Joe?" someone called out.

"I'm an eternal optimist, so I think we could do it, but I don't have the whole company to worry about," he replied.

"What about our study teammates, Honeywell and STL?" another asked.

"They'll have to join someone else's team also."

After some grousing we went back to our desks to consider our options. Gavin and Munier placed calls to some of the avowed prime bidders with whom we had prior discussions. Within a week Gavin and Titterton again addressed our group to announce a deal with General Electric: Grumman would support GE's bid and have major responsibility for the command module. We would all soon be temporarily assigned to a proposal headquarters that GE was setting up in Philadelphia.

A few days later, before we headed for Philadelphia, Gavin and Munier put me in charge of the Grumman technical proposal effort. Although disappointed, Sanial accepted this change graciously. Coming out of Munier's office, where Joe and Al had privately broken the news to him, he came over to my desk, shook my hand self-consciously, and promised his full support.

General Electric set up a large Apollo proposal headquarters on two floors at 30 Broad Street, in a modern office skyscraper in downtown Philadelphia, not far from Independence Hall and the Liberty Bell. Although the GE Space Division had moved to a new campus in suburban Valley Forge, they still had facilities downtown, including 30 Broad and the Chestnut Street building in which the division began, which gave them a convenient base from which to quickly staff and equip the proposal center. Logan Cowles, a stern, silver-haired veteran GE engineer, was Apollo program manager, reporting to the division president, Hilliard Paige. George R. Arthur was proposal manager, and Ladislaus W. "Lad" Warzecha his deputy; they had headed GE's funded Apollo feasibility study.

With the rest of Grumman's Apollo team, I moved into a hotel near the proposal headquarters and spent five or six days a week working there, returning home via Amtrak and the Long Island Railroad for one- or two-day weekends. Isolated from our homes and families, we poured all our energies into the proposal, often putting in fourteen- to sixteen-hour workdays.

The response deadline for the Apollo RFP was early October. GE led an intense, well-organized proposal effort, and the proposal headquarters itself was spacious and fully equipped. Each team member company had its own area, with partitioned cubicles and offices for the managers and supervisors and open "bullpen" areas for the workers. The attractive environment was well lighted, with window walls and recessed fluorescent lighting overhead, carpeted floors, and comfortable air conditioning—a major improvement over Grumman's PD rabbit warren back home. Near the GE office area were large conference rooms and an open area in which everyone could assemble for general meetings.

GE established a schedule of daily and weekly meetings that assured good communication among the proposal staff, which at its peak numbered more than six hundred people. Specific assignments were given, and the results were reviewed with technical management groups in the various specialties, selected to include all the participating companies.

GE introduced us to the use of "flip charts": large, easel-mounted pads of paper for developing and recording the ideas of brainstorming sessions and meetings. One session developed the idea that we could get "extra credit" and increase the depth of our proposal by conducting a comparative study of the various ways of performing the lunar landing mission. This was not required by NASA's RFP, which was largely silent on the mission approach to be used because NASA had not yet seriously confronted this choice.

The team members listed all possible mission approaches and the pros and cons of each. Each company team was asked to choose one of the approaches to study. I reviewed the options well into the night with Stern, Watson, and others and decided that lunar-orbit rendezvous—using a vehicle that could separate from the spacecraft in orbit around the Moon, land on the surface, and then launch itself back into orbit to join up with the spacecraft again—was the simplest way to get to the Moon and back. The next morning Gavin agreed, and we volunteered to conduct the LOR study for GE.

Erick Stern led a brief but useful study that produced a description of how the LOR mission could be performed, gave a rough cut of the size and weight of the major flight elements (booster and spacecraft), and outlined the major unknowns and technology hurdles. Most notable was our finding that the mission could be launched with a single Saturn 5, the huge rocket booster under development by Wernher von Braun and his team of German rocket engineers at NASA's Marshall Spaceflight Center (MSFC) in Huntsville, Alabama. The competing approaches all required at least two Saturn 5 launches or a much larger booster rocket, dubbed Nova. LOR's major risks were the lunar-orbit rendezvous maneuver itself, conducted far from Earth and ground-based guidance systems, and the development of a unique manned spacecraft optimized for lunar landing and liftoff.

The proposal was wrapped up in early October, at which point GE took over the final editing, publication, and delivery. Venturing forth from the office tower for an unaccustomed leisurely lunch in a nearby restaurant, I was surprised at the cool midday breeze. I had missed summer!

To help me unwind from work on the proposal, Joan and I enjoyed a long Columbus Day weekend on Nantucket. We had a brief but happy escape and discovered a wonderful island resort to which we returned many times with our young children. We liked to stroll down to nearby Brant's Point in the evening and watch the last ferry from the mainland make the tight turn around the lighthouse, its festive lights dancing on the dark water. Then I would look up at the Moon and say, "Look up there, kids. Your Daddy's going to build a spaceship to go there someday."

"Can we go too?" was the usual reply, accompanied by a frown from Joan, who knew we hadn't won the competition yet. That was my major worry also. Getting to the Moon should be easy compared with winning the contract, I thought.

On 28 November 1961 NASA announced that North American Aviation (NAA) was the Apollo spacecraft contractor. I collapsed into an emotional funk. In the drab PD mezzanine, the gloom was funereal. Sanial and I hung black crepe over the blackboard in our cubicle and wrote a note on it: "Thanks for all your help and hard work. We'll win the next one!"

There was not much more we could say. I was discouraged by the loss and felt that things might have been different if we had been prime contractor. Not that I could fault the way GE ran the proposal, but I did think they had assembled too large and complex a team.

A Second Chance

As the gloom subsided, Sanial, Munier, and I decided that our best remaining chance would be to take the LOR side dish from our work with GE and make it a main course: we would sell the LOR mission approach to NASA and then compete for the lunar landing spacecraft it required. The lunar lander was another exotic space flying machine that Grumman was as qualified as anyone else to produce. Gavin and Hedrick counseled that although this seemed to be a logical plan, perhaps we should seek NASA's advice directly. Grant Hedrick called his old friend Bob Gilruth, with whom he shared an interest in the design of high-speed hydrofoil watercraft, and set up a meeting for us with NASA's Space Task Group.

Early on a clear, cold Pearl Harbor Day we left the Bethpage airport in Grumman's G-1 propjet executive airplane, bound for NASA's airfield at the Langley Research Center in tidewater Virginia. The group included Gavin, Hedrick, Ferdman, Sanial, Watson, Stern, and others. We planned to present a summary of our latest study results on LOR and notional designs of the lunar lander. We spread out our material on the tables in the luxuriously appointed airplane and nervously reviewed the points we wanted to make. Unstated, but palpably present, was the realization that this was our last chance at Apollo.

We were warmly greeted by Bob Gilruth and Max Faget. A number of STG engineers, including Bob Piland, Owen Maynard, and Caldwell "Cadwell" Johnson, gathered to hear our briefings. The briefings went on all morning, with questioning and discussion by the NASA people and steadily growing attendance. I felt that we made a strong case for LOR and showed good understanding of the principal problems and tradeoffs involved in the design of the lunar lander. When we spilled out into the hallway at the lunch break, I overheard Gilruth tell Faget, "I think you should show them everything we've done."

After lunch in the building's cafeteria, Faget told us they were impressed with our results, which seemed to match NASA's own internal study findings.

He proposed that they show us their results and compare them with ours. For the next three hours we held detailed engineer-to-engineer discussions on the merits of LOR versus the competing approaches: direct ascent and Earth-orbit rendezvous (EOR).

We also looked at NASA's designs for the lander, which they were calling the lunar excursion module, or LEM.[4] Of particular interest were comparisons of configurations and equipment arrangements and weight estimates for the NASA and Grumman versions. There were novel features of each group's designs, but the total weight estimates were approximately the same. Gilruth and Faget said that they believed LOR was the approach NASA should select for the Apollo mission, although this opinion was still not prevalent among NASA management.[5] They thought we had made a good start with our LM designs and encouraged us to pursue them further in anticipation of a possible industry competition.

On the return flight we were buzzing with excitement. Hope was reborn! Gavin asked me to prepare a study plan and budget request for the next year, aimed at positioning us to bid as prime contractor on the LM. I bounded off the plane in Bethpage refreshed with the prospect of a marvelous new challenge.

The twelfth of December was a miserable morning, with a raw, gusty wind driving a steady rain. I felt chilled just walking from the parking lot into the front lobby of Plant 5, and I climbed up the stairs shivering. I had decided to attend the weekly meeting of the Propulsion Section on the main Engineering floor. It was a way to keep in touch with my other Grumman "home," a good idea with my prospects in space looking so shaky.

I was early, alone in the Propulsion area, just a few people streaming across the main bullpen. As I sat debating whether to go downstairs and pick up some coffee, a telephone at a nearby desk rang insistently.

A barely audible woman's voice said, "Hello, is this . . . is this Grumman?"

I felt the hairs on the back of my neck rise.

"I just wanted you to know . . . to know that Tom Sanial was killed . . . in an auto accident this morning."

"What? What? Who is this?"

She was sobbing. "A neighbor. Just a neighbor. I'm so sorry. So very sorry. I must call others now. Goodbye."

I stared unseeing at the telephone in my hands. A few minutes later I walked to Joe Gavin's office, blurted out the gist of the call, and burst into tears myself.

Sanial's loss deprived us of his extensive Apollo knowledge and background at a critical time in Grumman's efforts. Others in our study team rose to provide the systems design integration talent for which we had relied on Sanial. The cruel but practical lesson was that in a big company, no one is indispensable.

On every raw, rainy December day, I think of the immense promise and goodness that was snuffed out so senselessly, and wonder at the implacable role that fate and chance plays in our lives. And I miss that sweet, gentle man who surely would have shared with me one of mankind's greatest adventures.

Getting into Position

The company authorized fifty people to study LOR and the LM for a year. Joe Gavin headed the project, and I led the technical study. We received major assistance from the Radio Corporation of America (RCA), which provided a sizable engineering team to support our studies at their own expense. RCA was responsible for most of the electronics and for some of the systems engineering. Their team was headed by Frank Gardiner, a darkly handsome, smooth-talking senior electronics engineer. About thirty RCA engineers moved into the PD mezzanine with us.

Gardiner was able to tap experts from different RCA divisions to bring the talent we needed to the team, including communications engineers from Camden, radar specialists from Burlington and Moorestown, and guidance and control experts from Burlington. They bolstered our effort with in-depth technical expertise and marketing savvy.

In January 1962 we competed for a NASA-funded study of LOR and the LM. Although we thought our proposal was a good one, Convair won the award: fifty thousand dollars for a four-month study. We proceeded with our company-funded study anyway, and in June we submitted our study report to NASA. Shortly thereafter we were invited to brief our findings to Joseph F. Shea at NASA Headquarters in Washington, D.C.

Shea had recently been recruited to NASA by Brainerd Holmes, NASA's associate administrator for Manned Space Flight, and had been assigned to settle the "mission mode" issue. An experienced systems engineer from the Titan ballistic missile guidance program, Shea projected intelligence, engineering talent, self-confidence, and leadership. He was the right man to make a momentous decision.

In my first meeting with him in Washington, Shea continually interrupted my briefing with difficult but logical questions and meaningful comments: What makes you so sure the rendezvous can be accomplished? It's a long way from home, and there won't be much help from the ground. Have you calculated the allowable guidance errors for each rocket firing during rendezvous? How good are your LM weight estimates? If LM is overweight, it gets multiplied all the way down the launch stack.

Our study results on the relative advantages of LOR were by then quite mature, and I was on solid ground with our data, able to parry Shea's thrusts. Our LM design studies had also progressed to the point where they seemed credible, and each major design feature was supported by technically satisfying arguments. John Houboult joined enthusiastically in the interrogation; it was like defending a doctoral dissertation.

After two hours of grilling, Shea smiled and said that we had done a useful study on our own initiative and promised to consider our input in reaching his decision. He complimented me on my presentation and in-depth knowledge. I left the room elated that I had survived a baptism by fire.

Two weeks later NASA announced that they had selected LOR as the Apollo mission mode and would proceed with an industry competition for the design, development, and construction of the lunar module. The LM request for proposal was issued in late July, with responses due in early September. We were ready. After more than three years of preparation, Grumman was in the right place at the right time. And I was hungry for a win.

3

The LM Proposal

From the praise and comments of our Grumman supervisors, we knew that we were an elite within the company, chosen from among the brightest in a demanding profession in which brainpower ruled, counted upon to create the systems that would become the mainstays of the company's business. We had been assigned to Preliminary Design, the nest from which new airplanes were hatched to fly "higher, faster, and farther." Now this quest had reached its ultimate conclusion: escape from Earth itself and flight to our nearest celestial neighbor. At thirty-two years of age, I was leading Grumman's technical proposal to NASA to design and build a spacecraft to carry men to the Moon and back.

To the uninformed observer, however, we looked like outcasts confined to a hidden backwater where we could do no harm to the company's ongoing business. We worked in a segregated area suspended from the ceiling over a portion of the Experimental Shop, reached by a nondescript flight of dark blue painted metal stairs leading up from the polished wood blocks of the shop floor or by a flight of metal stairs and a catwalk down from an unlabeled door on the second floor Engineering Department office. Both stairways led to a blue metal door in a whitewashed cinder-block wall to which a doorbell and buzzer provided access. Inside, the low-ceilinged compound bathed in fluorescent light and humming with air conditioning seemed like a time tunnel, remote from worldly existence.

Our office area was cinder-block painted a faded light yellow and crammed with as many wooden desks and chairs as would fit. A single small office at the front of the room, framed by a large window partition, belonged to Al Munier, one of the few permanent members of Preliminary Design. Al was an experienced aircraft designer who had helped fashion Grumman's Wildcats and Hellcats during World War II and a firm proponent of advancing the company into the space age.

The group secretary's voice crackled over the intercom, summoning me to a meeting in Joe Gavin's office. I put aside the study report I was reviewing and bounded up the stairs to the second floor. It was a welcome change to be on the Engineering floor. The interior vistas were broader and the ceilings higher and neatly finished with white acoustic tiles and frosted glass panes concealing fluorescent light bulbs. Joe Gavin's office was spacious by Grumman standards. The walls were tastefully covered in rich dark paneling, and it was furnished with dark mahogany furniture with brass hardware and trim.

Joe and Al Munier were already seated at the conference table, and I joined them there, followed by my deputy Erick Stern. "The LM RFP has been released," said Joe in his crisp voice with a hint of New England twang. "Saul Ferdman just picked it up in Houston and we'll have it here in the morning. The proposal is limited to one hundred pages and it's due in sixty days. Al and I thought we should get together and make sure you have everything you'll need."

Joe Gavin, in his early forties, was a rising star at Grumman. Joining the company in 1946, with an aeronautical engineering degree from MIT and wartime service in the navy's Bureau of Aeronautics, he soon established a reputation as a talented aircraft designer with leadership capability. As project engineer on the swept-wing F9F-6 Cougar, he directed the design of an improved stabilizer control actuator driven by a high-speed, irreversible, ball-bearing screw jack. This novel design allowed the Cougar to safely fly through the Mach 1.0 speed-of-sound barrier in a dive. The Cougar and its straight-wing predecessor, the F9F-5 Panther, both served in the Korean War. Joe became project engineer on the F11F Tiger, the first Grumman-produced fighter that was supersonic in level flight. The Tiger reached limited production and contained many technological innovations.

Joe Gavin grew up in eastern Massachusetts near Boston. His father was a tinkerer and tool collector, and young Joe developed a keen curiosity about how things work. A visit as an eight year old to a local dirt airstrip where he saw the transatlantic hero Charles Lindbergh impressed him early with the romance of aviation, and his studies at MIT informed him of its technical elegance. By nature reserved and understated, he thought problems through and stuck with his conclusions. Tall, muscular, and with craggy good looks, he was an accomplished oarsman at MIT and an expert skier from his youth. He was a natural leader who, in the face of crises and confusion, remained calm and steadfast of purpose, inspiring others to rally around him.

Gavin was chief Missiles and Space engineer, in charge of the LM project, and my boss. Al Munier provided the proposal team with support and guidance from the resident Preliminary Design Group. We went over a list of additional engineers we would need and added office space and equipment. High on our equipment list were IBM Selectric typewriters—the new design with the removable type ball—which were in short supply at Grumman.

Most of the engineers we needed had worked on our proposal for the Apollo spacecraft in 1961 and on LM studies since then.

Al, Erick, and I assembled the dozen or so LM proposal engineers in our work area and passed the word. Al promised to find space and desks for another two dozen people in Preliminary Design; the remainder would have to work at their "home" desks and visit us as required. A large conference room in the mezzanine would be available for our daily proposal meetings, which would begin as soon as we had the request for proposal.

The next morning Joe, Al, Erick, and I joined a standing-room-only crowd in the plainly furnished conference area within Preliminary Design. Saul Ferdman, Grumman's space marketing director, passed out copies of the RFP and summarized what it contained, having read it on the plane from Houston. It was a drastic departure from NASA's usual RFP, which normally provided a detailed set of mission plans, spacecraft specifications, and technical requirements and requested that contractors respond with their preliminary design of spacecraft and systems to meet the requirements, their plans for building and supporting the spacecraft to a NASA-specified schedule, and their bid price.

For the lunar module NASA considered both the mission planning and the technical requirements too uncertain to buy a proposed design. Instead they decided to base contractor selection on an evaluation of which company's design team was most knowledgeable about the LM's mission and requirements and had plausible approaches to its design. The company's manufacturing capability, financial stability, and record of quality would also be considered, and the estimated cost for the program was requested as a means of determining the contractor's understanding of the program's scope.

The RFP was more like a graduate examination in an aerospace engineering design course than a typical government procurement specification. It posed fourteen technical questions and required discussion of five management areas, to be answered in one hundred pages of carefully specified format, even to the type size and line spacing. The technical questions "probed the most exacting technical requirements in the LM mission," as we told NASA in our response.[1] Summarizing some of them:

1. Discuss the flight mechanics and other considerations of near-Moon trajectories and of lunar launch and rendezvous.
2. Describe your approach to the design of the following LM systems: onboard checkout, propulsion, reaction control, flight control.
3. To what extent do you consider backup methods of control and guidance necessary? Describe your approach to this issue.
4. How do visibility requirements affect LM operations and design?
5. How would you accommodate micrometeoroid and radiation hazards in the LM design?

The RFP encouraged contractors to submit a conceptual design of a lunar module with their proposals, as a means of focusing their answers to the questions and demonstrating their competence in manned spacecraft design. However, NASA was not buying the contractor's design; after the winner of the competition was selected, NASA and the contractor's engineers together would develop the preliminary design of the LM.

We already had three conceptual designs prepared against our own estimates of the mission requirements and the space environment. Our major tasks were to compare NASA's official requirements with our estimates, to determine the impact of differences on our designs, to select a leading candidate design, and to refine and improve that design until time ran out for the proposal. In parallel we drafted answers to NASA's questions and analyzed them for possible effects on our conceptual design. Within a couple of days we were deeply immersed in this process.

We ultimately submitted a design for a two-part spacecraft, with a lower landing, or descent, stage and an upper liftoff, or ascent, stage. The descent stage contained mainly the tanks, rocket engine, and plumbing of the descent propulsion system, which was used to drop the LM out of lunar orbit and land it gently on the Moon's surface; other consumables, such as oxygen, water, and batteries, which could be left behind on the Moon; scientific equipment to be deployed by the astronauts during exploration; and the landing gear.

The ascent stage contained the crew compartment and cockpit in which the two astronauts who flew the spacecraft between the Moon's surface and lunar orbit lived, ate, and slept while on the Moon. This stage contained most of the electronics systems: GNC; communications, radar, and instrumentation; the reaction control system (RCS), with its sixteen small rocket engines that allowed the pilots to control and maneuver the LM in flight; and the environmental control system (ECS), which provided conditioned oxygen and water to the crew for life support and the spacesuits and backpacks needed to go outside the LM's pressurized cabin. The crew compartment had two access hatches: one on top for docking with, and crew transfer into, the command module, and one forward for crew egress to the lunar surface via a platform. An external docking module allowed either hatch to dock with the CM as a redundancy provision.

The crew compartment was modeled after that of helicopters, with the astronauts seated in a large forward-facing glass bubble with an instrument panel between them. The landing gear had five fixed legs that just fit within the spacecraft/LM adapter (SLA), inside which the LM was housed at launch from Cape Canaveral. This landing-gear design was what yacht racers would call a "rule beater": it barely satisfied the critical tip-over and surface-penetration requirements of the RFP without the added complexity of an extendable landing-gear mechanism, its five legs providing enough tread width and pad area within the SLA space envelope to make this possible. Even as we sub-

The lunar module proposal design. (Courtesy Northrop/Grumman Corporation)

mitted the proposal, I did not expect the real LM to have a fixed landing gear because the slightest change in design assumptions or LM weight would negate this approach, but it was simple and lightweight for the proposal.

The conceptual design met all of NASA's performance requirements and weighed only twenty-two thousand pounds fully loaded, well below the RFP limit of twenty-six thousand pounds. Grumman's model shop, accustomed to making beautifully lacquered and detailed display models of airplanes, did the best they could with our ugly duckling, but it remained an alien machine suited to other worlds. The nickname "Bug," which NASA often used in their studies, still seemed to fit our creation.

The last of NASA's technical questions was a zinger: What are the five most important considerations in the design of the LM? List in order of descending importance and explain your reasons for selection.

I sensed this was a make-or-break question and kept revising our answer until the end. With my technical group I drew up lists of candidate issues and the reasons they were important. Stern, Watson, Gardiner, and I reviewed and debated these and sought opinions widely, from Gavin, Munier, and Ferdman, and from key players, such as Bob Mullaney, Bill Rathke, Bob Carbee, and Arnold Whitaker, who were named in the proposed project organization to join us if we won. At the printer's deadline, Stern and I finalized the list:

1. Propulsion design and development
2. Flight control system design and development
3. Reliability
4. Weight control
5. LM configuration[2]

We hoped this would agree with the list Max Faget and his designers had in their minds. Recalling our first meeting on LM with Faget and his group at Langley Research Center, I felt confident that our similar engineering approach and reasoning would lead us to the same answer.

In the program management section of the proposal, NASA asked for the company's related experience and performance, our proposed LM program organization and its relationships to the rest of Grumman, our facility and manpower capability, a make or buy plan, and cost estimates. Gavin and Ferdman ran this part of the proposal, but my engineers and I were heavily involved, as our requirements and plans affected everyone else. For example, one of the technical questions asked us to explain our test and development program plans. These established a far-flung series of LM test-beds, test facilities, and qualification programs that consumed a major portion of the time, money, and manpower needed to complete the LMs. Engineers provided the manpower estimates for the design and test phases of the program and part of the manufacturing phase. Making the cost numbers add up and agree with our descriptions of program content required many iterations.

An aerospace proposal team works as hard as they can to produce the best possible proposal until time runs out. For two months our team of about eighty engineers discussed, debated, analyzed, and wrote our responses to NASA's exam, cramming as much as we could into one hundred pages of print and illustrations. We included a large foldout general-arrangement drawing of our LM design that showed many of its components, complex functional block diagrams of systems, and a large flow diagram of the development plan. We selected subcontractors and suppliers for LM's subsystems and major components after holding minicompetitions, and their key people helped with our proposal. The whole team routinely worked fourteen to sixteen hours a day, seven days a week, relentlessly driven by the belief that our competitors were working at least as long and hard as we were and our only chance of winning was to give it everything we had. Like an eight-oared crew, we did not want to have anything left when we crossed the finish line.

Inside our windowless warren, we lost track of time. One evening when our full crew was hard at work, three maintenance men intruded into our space carrying large stepladders, pipe wrenches, and other tools. Without a word to any of us, they set up the ladders and started dismantling the sprinkler system over our heads. When flakes of rust and then a thin stream of water showered down on the pages I was reviewing, I sprang up to protest. The workmen retreated in the face of my determined objections, but not before telling me who their boss was and assuring me that, whether or not I liked it, the sprinkler system work had to be done.

The next morning I complained to Al Munier and asked him to get the sprinkler work stopped until work on the proposal had been completed. Returning to the day's activities, I forgot all about the matter. Late that afternoon, however, Al called me into his office. He upbraided me for being rude to the workmen and said that if I worked more efficiently I would not have to be in the office in the middle of the night. I started to object but quickly stopped. Looking at Al's face I saw the warning signs of an impending temper tantrum.

When Al's temper took hold, his face flushed, his eyes narrowed, and his lips compressed into a hard, thin line. He accused me of arrogance, insensitivity, and inefficiency and told me the sprinkler system had to go in, whether or not I liked it, and while the workmen were there they would clean the light fixtures and paint the ceiling too. Then he abruptly ordered me out of his office.

Every evening for the next three weeks we faced a daunting succession of distractions and interference. First the plumbers installed new overhead sprinkler pipes, then the cleaners and electricians dusted and wiped the overhead fluorescent light fixtures, adding new fixtures and replacing burned-out bulbs. Then the painters, with their long rollers, white caps, and paint-spattered coveralls, worked on the walls and ceilings. All this sent debris showering down on the desks and floors of our area. It was comical, considering all the work we had to do and our need to concentrate on imagining the design of a unique spacecraft. We survived by packing up our papers and drawings when they appeared, spreading drop cloths over our clean desks, and borrowing desks in an undisturbed area on the far side of the mezzanine. In the brief moments I had time to think, I wondered whether the company understood the importance of winning this contract.

Two days before the submittal deadline, we turned the proposal over to the document production process. Erick and I gave the galley proofs and graphics a final end-to-end read-through and handed it all to the proposal editor, who took it to the nearby outside print shop for final makeup and printing. Then we joined the rest of the proposal team for a celebratory lunch. I had reluctantly agreed to the team's request to schedule the lunch before the proposal was actually delivered to NASA because the Labor Day weekend was approaching and many of our people wanted to stretch it by taking off the preceding Friday. After all the long hours they had put in, I did not want to appear ungrateful. But when I walked into the restaurant and saw our people enjoying cocktails, my stomach knotted up. What if we all suddenly had to go back to work? I never liked having alcohol anywhere near the job.

I had been in the restaurant about twenty minutes when a waiter told me I had a phone call. It was Saul at the printers. There were two problems: first, when the final page layouts were completed, the book was almost a whole page too long. Something had to be cut. And second, as Saul was reading through the proofs, he discovered an inconsistency between our answers to two of the questions. Which one was correct?

I thought Saul and I could fix the problems ourselves, and so I quickly headed for the door, but Erick saw me leaving and followed. The printer was in a squat gray building just outside the fence from our area of the Grumman complex. We found Saul with our editor and the printer's project leader poring over page layouts. Erick and I joined them, and after about an hour we had marked up enough snippets for deletion to bring the proposal within one hundred pages. We were unable to resolve the inconsistency, however, and were forced to call back three experts from the luncheon. We discovered several other statements in our rereading that seemed questionable and had to be verified with the people who wrote them. Before long we had about ten people in the printers, trying mightily to focus on minute details of the proposal after coming from what had developed into a roaring party. To their credit, they were soon able to satisfy me and Erick about their sections or make minor modifications for added clarity, but it was a frantic couple of hours before all the loose ends were secured and the proposal was again declared finished.

The next evening I was finishing dinner with Joan and the kids when a call came in from Saul. He had safely delivered our proposal two hours before the deadline. Hallelujah! At last it was over. I looked forward to taking the next two weeks off, using some long-deferred vacation time. After a few days at home, Joan and I would get away by ourselves for a week in Williamsburg, Virginia, and the Blue Ridge Mountains while her parents stayed in our house with the children. Sleeping late and seeing the outside world again after being sealed in the time tunnel. I could hardly wait.

4

The Fat Lady Sings

Three weeks after the LM proposal was submitted I flew to Houston on the luxurious Grumman Gulfstream with the LM program team and some of the company's senior executives to brief NASA on the "salient features" of our proposal and to answer their questions. The briefing, "orals" before an invisible audience in a darkened auditorium, were an ordeal, but thanks to thorough rehearsals we acquitted ourselves well.

On the trip down and back I found that Clint Towl and Bill Schwendler were almost as excited about going to the Moon as I was. In an expansive moment on the return flight, Schwendler promised to fly us down for dinner at Antoine's, the renowned New Orleans restaurant, if we won the contract. Towl was aghast and assured us that this was so counter to Schwendler's conservative nature that he would never follow through, even if we did win. Towl knew his friend and colleague well—no more was ever said about Bill Schwendler's promise.

About two weeks after the orals our administrative planning was abruptly halted by a phone call from Bob Mullaney summoning me and other project leaders to drop everything for a "fire drill," our expression for a sudden emergency task that must take priority over all others. (Now more smoothly referred to by management consultants as "crisis management.")

The fire drill was a list of questions from the NASA Source Evaluation Board that had to be answered within one day. It was not just an exercise but an attempt by NASA to gain additional information they needed in evaluating our proposal. They were serious questions, probing our plans and capability for doing the work and our interpretation of certain technical requirements. Ferdman got the deadline extended, and at his urging, Gavin decided to assemble a small group and deliver the answers in person. In less than forty-eight hours we wrote a thirty-page miniproposal and handed it to an astonished Bob Piland, head of the NASA evaluation team in Houston. We had the opportunity to discuss some of our answers with Piland and restate the main themes of our proposal. On the flight back to New York we were giddy at the thought that an incredible adventure appeared almost within our reach, and we fantasized about its possibilities. Mullaney, Ferdman, and I fairly vibrated in our seats with suppressed excitement.

After the orals and the fire drill, it was difficult to get back into the routine work of planning how we would perform the job if we won. Rumors swirled daily, alternately tantalizing us and dashing our hopes. We tried to ignore them, telling one another, "It's not over until the fat lady sings." But it was no use; I for one could not concentrate without drifting into fantasy, imagining how wonderful it would be to win this prize sought by all the giants of aerospace, and to design and build a manned spaceship to land on the Moon. What would it look like, and how could we ever design and build it in time? My colleagues and I were suspended in a time warp of anticipation, drifting from rumor to rumor in a fuzzy haze, believing that things would turn out well for us.

The rumors of the prior week focused on election day, Tuesday, 6 November 1962. Some said the LM contract would be awarded before the election to gain favor with voters in the selected state and region; others maintained it would be delayed until after the election to avoid disappointing the losers. Election day arrived with still no word, so half the rumor was proven false.

At eight o'clock the next morning I was back in my Preliminary Design cubicle looking over the material I needed for the day's planning effort. I had enough information on staffing and space requirements for the first six months after go-ahead to meet with LM program administrator Art Gross and firm up floor plans and building space commitments. Because our staffing buildup assumed a 1 December 1962 go-ahead, we were initially confined to space on the main Engineering floor in Plant 5. When the new Plant 25 Space Engineering Center opened in February 1963 we would have plenty of room for the first two years' worth of growth, but the transition from our start-up in Plant 5 to permanent quarters in Plant 25 required careful planning to avoid breaking the momentum of program activity and growth.[1]

I was waiting for Art Gross to arrive for our meeting when Erick Stern, grinning from ear to ear, entered my office, stuck out his hand, and said, "Congratulations!"

"For what?" I replied.

"We won! I just heard."

About that time my phone rang; it was Joe Gavin. "I've just been informed that Clint Towl will be receiving a call from our local congressman in a few minutes," he said. "It's not quite official yet, but it sounds like we've won! Congratulations! I'll call back when it's official."

I had no sooner relayed Joe's message to Erick and others who had gathered around my desk with excited, smiling faces when Joe Gavin called again.

It was official: we won! Representative John Wydler (R-N.Y.), Sixth District, called Clint Towl and read him the official announcement press release, which was being sent to Grumman immediately. Joe thanked me for everything I had done to make victory possible and asked me to extend his appreciation to the entire LM Engineering team. Even before I finished my conversation with Joe, my broad smile and "thumbs up" told the story to the growing crowd around my desk, and a rousing cheer shook the dusty corners of the PD mezzanine. I circled through the area shaking hands and slapping backs, hardly daring to believe the news.

The next few days were a blur of activity and excitement. There was an "all hands" meeting with our corporate leaders in which Clint Towl, Bill Schwendler, George Titterton, and Joe Gavin thanked the LM project team for their successful efforts on the proposal and promised help and support for the challenging times ahead. They hosted a celebratory lunch at the Grumman executive table at the nearby Beau Sejour[2] restaurant for the LM project, where toasts were drunk in ginger ale (no alcohol, in the Grumman tradition that work and drink did not mix), and the corporate leaders expressed their intense pride and pleasure that Grumman would be part of the ongoing national adventure. Bill Schwendler came close to making a speech, offering his congratulations and promising support as needed.

There was a frantic after-hours party for the LM proposal team at the nearby Holiday Manor catering hall, where plenty of alcohol fueled a wild victory celebration. During the day there were dozens of congratulatory phone calls from the NASA people we knew, our RCA teammates, and other subcontractors who had supported our proposal. Norm Ryker called me from North American, the prime contractor for the Apollo command-and-service module spacecraft, welcoming Grumman to the Apollo team. We had arrived into the elite Apollo fraternity.

NASA asked us to come to Houston in mid-November to begin contract negotiations. We needed to bring all our proposal backup data and our experts in every area of the program. The negotiations were expected to take several weeks. We put Art Gross in charge of the negotiation team logistics, and he took over in his authoritative fashion. Within a few days we identified a Grumman negotiation team of about eighty people, and Gross supplied them with airline tickets, rental cars (car-pooled four to one), cash advances, and hotel reservations. Someone thought to ask the hotels if the fact that two of our people were black posed any problem, and in the Houston of 1962, it did. One of the hotels, not wanting to lose our business, said that our black members could reserve a room provided they avoided the lobby and the dining areas. The hotel said that, if necessary, it could "fix them something to eat out in the kitchen" (!). Gross phoned several hotels and motels, progressively more distant from our NASA destination, until he found one that was color blind: the Sheraton Lincoln Hotel in downtown Houston, into which he booked our whole team.

NASA had its own logistics problems with the LM negotiations. The campus of the Manned Spacecraft Center at Clear Lake was still under construction and would not be ready for several more months, and the barrackslike World War II buildings at Ellington Field were filled with the burgeoning NASA Apollo program staff. All available speculative office space in southeast Houston had already been leased by NASA, so they had to be ingenious to find more.

NASA located a real estate developer who was building a large garden apartment complex off the Gulf Freeway in southeast Houston, next to the just completed Gulfgate Shopping Mall. They leased two apartment buildings that were completed structurally but not finished on the interior and had the developer make them into temporary offices. They had water, electricity, air conditioning and heating, telephones, bathrooms, and basic interior white paint but lacked carpets and flooring, kitchens, and many of the doors, fixtures, and decorations. These finishing touches to make the buildings into attractive apartments would be added later, after their use as a makeshift office complex had ended. Openings were cut into some of the walls between apartments to permit interior access and circulation throughout the building, and the parking area adjacent to the buildings was temporarily paved with crushed oyster shells from the Gulf of Mexico. A permanent asphalt surface would come later, after the traffic of heavy construction equipment subsided.

We were greeted at Gulfgate Gardens by Charles Frick, Apollo spacecraft program manager, and many other members of the NASA LM Negotiation Team. Gulfgate Gardens was a large, not unattractive complex of two-story red-brick residential apartments, with white architectural trim and white shutters on the windows. The buildings contained ten or twelve duplex units, each with a separate front door, and were widely spaced, surrounded by lawns and young, sticklike trees and shrubbery, all newly planted.

NASA had two adjacent buildings on a quiet street in the interior of the complex. When we arrived, furniture trucks were parked outside the NASA buildings and workers were busily unloading the desks, chairs, and file cabinets NASA had rented. Inside was mostly antiseptic white, but there were clumps of construction debris in the corners here and there. Wires dangling from ceiling electrical boxes and covers missing from wall switches and plugs emphasized the unfinished nature of the place. The rooms had been assigned to some thirty specialty teams that would fact-find and negotiate each area of the proposal. The NASA and Grumman team members paired off and got acquainted in their new quarters.

I joined Gavin, Mullaney, and Rathke in a meeting with the NASA leadership. NASA said they wanted to review our backup data showing the detailed basis for our proposal, including technical approach; program, manufacturing, test, logistics, and staffing plans; subcontract plans; and manpower, cost, and schedule estimates. After this joint fact finding we would discuss changes to the proposal, which would become the negotiated basis of the LM contract. NASA thought this process should take about two weeks and targeted completion the day before Thanksgiving. They asked, however, that we plan to remain in Houston over Thanksgiving and into December in case the negotiations took longer than expected. We agreed to a weekday schedule with counterpart meetings beginning at 10:00 A.M. and continuing nominally until 5:00 P.M.; this allowed each side to hold summary status meetings beginning at 7:00 or 8:00 A.M. and use the evenings when necessary for continued counterpart discussions and for separate problem resolution and status accounting on each side. Saturdays would be used as necessary; Sundays would be days off. Before launching into this demanding regimen, NASA invited the combined negotiating teams to a "kick-off" luncheon at a nearby Gulfgate restaurant.

The NASA LM Negotiation Team was led by Bill Rector, LM program manager for the Apollo Spacecraft Program Office (ASPO), with Dave Lang, MSC contracts director, as the team deputy. Rector had only been with NASA a few months, having been recruited by Charles Frick, his former boss at General Dynamics, shortly after Frick joined NASA in January 1962. Rector had been a key member of General Dynamic's Apollo spacecraft proposal team in 1961, and wanting to be part of the program even though his company had lost, he decided to join Apollo with NASA. I had met him at Apollo bidder's conferences and technical society meetings during the prior two years. He was a friendly, direct fellow with an aerospace engineer's understanding of technical and program details and the intricacies of doing business with the federal government. I felt a basic rapport with him as one who empathized with the contractor's problems and concerns. Rector was a tall man with wavy brown hair, a boyish-looking freckled face, a snub nose, and wide, round eyes, which he hid behind owlish horn-rimmed glasses. I thought he looked younger than me, but we were probably about the same age. Because he had only been with NASA a short time, Rector relied heavily on Lang and other NASA veterans for advice, but he was quite capable of making his own decisions. Major agreements and the overall negotiated contract would have to be approved by Bob Gilruth, MSC director, and Frick, as well as by George Mueller, Apollo program director at NASA Headquarters, and his staff.

At lunch we relaxed with the NASA people with whom we would be intensely involved for years to come. We learned a little bit about them as individuals, and about how they reacted to Grumman's LM proposal. In general they thought our proposal was ingenious and showed great knowledge of the mission and its technical problems but was naïve and simplistic in a number of areas. It promised to be an interesting negotiation.

Bob Carbee, Arnold Whitaker, Erick Stern, and I at first paired off with

Owen Maynard and his chief assistants to review the overall technical approach. NASA explained to us that the LM's requirements, mission plan, and space environment would continue to evolve as NASA and Grumman worked together closely for the next several months leading to the LM mockup review. NASA, therefore, was not buying the preliminary LM design that Grumman proposed but planned to develop the LM's design together with Grumman. NASA would provide direction and leadership and share their knowledge of space engineering; Grumman was responsible for developing, implementing, and testing the complete detailed design. True responsibility for the LM's success or failure would be shared between NASA and Grumman and would depend upon the cooperative efforts and capabilities of both. But for contractual purposes, Grumman's tasks would be clearly defined by technical performance specifications, cost targets, and schedules. Our task in the negotiations was to envision how the LM was likely to evolve in the next year, to rough out a revised technical approach, a program plan and cost and schedule estimates based upon that vision, and to use this as the basis of the initial LM contract.

My colleagues and I had not thought of these negotiations in such specific terms before, and initially we followed NASA's lead. For example, Owen Maynard questioned our fixed, five-legged landing gear. NASA considered it inevitable that the LM's weight would increase substantially, requiring a greater footprint width of the landing gear that would necessitate a folding, deployable gear. They suggested that our technical approach be changed and cost and schedule estimates increased to account for the design and development of the more complex deployable landing gear. My proposal "rule beater" was very short-lived.

We entered into the spirit of things, and before long the Grumman people were questioning NASA about the validity of some of our own proposal assumptions. After a few days everything in the proposal was up for grabs as we jointly took on the challenge of guessing what the LM design and program plan would look like after a year or more of intensive development. It was an unusual negotiation for a government agency and a contractor—repeatedly the government pointed out where Grumman's estimates were incomplete or oversimplified and the more likely approach would be more complex and expensive. Contrary to the usual practice, NASA was negotiating the price up, not down!

At our morning summary meetings we learned that these upward negotiations were occurring throughout the program. In some areas, such as ground-support equipment and logistics, NASA considered that we had grossly underestimated the scope of the job and ignored major areas of required GSE and logistic support. They also realized that we did not understand how NASA conducted factory and development testing on the Apollo program. We had little knowledge, for example, of the functioning and requirements of the automated checkout equipment (ACE), which was developed by General Electric and would be installed and operated by them at Grumman's factory in Bethpage and in the LM checkout area at Kennedy Space Center. The rigor and discipline with which NASA documented every action on the factory floor during spacecraft assembly and checkout also was beyond our ken. We never imagined that NASA required written step-by-step procedures, witnessed by an inspector, whenever a spacecraft or any flight hardware was touched, as in this procedure for attaching a support clamp to tubing:

> Step 45. Place support clamp P/N AN269972 on water line P/N LDW 390-22173-3 at location shown in sketch.
>
> Step 46. Verify that rubber grommet on clamp is properly seated with no metal touching the tubing.
>
> Step 47. Align holes on clamp with hole in structure P/N LDW 270-13994-1. . . .

It was inevitable that requirements such as this, which we dimly perceived after NASA pointed them out, would have a major impact on cost and schedule estimates.

As the negotiations proceeded the scope of the LM job was rapidly growing. Using their experience in manned spaceflight with Projects Mercury, Gemini, and Apollo, NASA was very helpful in identifying areas of underestimation in our proposal, and in visualizing the form in which growth might occur. There were other areas, however, in which even NASA had limited insight: such as the nature of the lunar surface, the difficulty of cislunar navigation and lunar landing, the limits of astronaut performance on the Moon, and the requirements of the space and lunar environments. In such cases our discussions produced consensus speculation, inevitably more complicated and difficult for the LM.

It became obvious that we were not going to finish by the Thanksgiving deadline. There were simply too many areas of the proposal to be redefined and reestimated. We told our team members to prepare for a longer haul; with luck and hard work we would be home before Christmas. As gently as possible we broke the news to our loved ones back home.

I was getting to know some of the NASA leaders with whom I would be working closely for the next few years. As NASA's LM Engineering manager, Owen Maynard was my direct counterpart. He was a young engineer from Canada, about my age, with a friendly, outgoing personality and an understated, dry sense of humor that made him a pleasure to deal with. Maynard was of medium height and build, with brown hair and narrow squinting eyes that often showed a mischievous twinkle. After becoming accustomed to the soft southern accents or southwestern drawls of most of the NASA people, Maynard's clipped Canadian speech sounded very different. His sentences were frequently punctuated with "Eh?" (rising inflection), which was not a question but a pause or acknowledgment, roughly equivalent to a New

Yorker's "uh huh" or "y'know." The words "oot" (out) and "aboot" (about) also signaled his Canadian background. Of course, Maynard thought that the Grumman "Noo Yawkers" spoke with strange accents, too, which he dryly let us know when we were rude enough to comment on his speech patterns.

Not only his pleasant personality but also his experience and technical background made working with Maynard rewarding. He had worked on aircraft design at A. V. Roe in Canada before coming to the United States and joining NASA at Langley, Virginia. He was one of the early members of Bob Gilruth's Space Task Group, working with Caldwell Johnson under the brilliant and innovative Max Faget. Maynard and Johnson had performed most of STG's in-house studies of alternate lunar landing mission concepts and were early disciples of John Houbolt's LOR approach. They developed STG's own preliminary LM design, which the LM Source Evaluation Board used as a standard against which to evaluate the contractors' design proposals. During the negotiations Maynard refused to show us the group's LM design because he thought we might be unduly influenced by it and our own inventiveness would be suppressed. I thought that differences between NASA's in-house LM design and the Grumman proposal provided the basis for many of the technical questions and concerns that NASA raised in our discussions.

Dave Lang, MSC contracts director, was also pleasant to deal with. He was a tall man with a craggy face, jet black hair slicked back, bushy black eyebrows, and the wise and wary look of a professional poker player. He favored cowboy boots, bolo ties, and a white Stetson hat, which, combined with his Texas-Oklahoma drawl, made him seem to me the quintessential southwesterner. Although he exuded an air of shrewdness, he was so friendly and helpful that it was impossible not to like him. When called in to mediate a disagreement between Grumman and NASA negotiators, he sought a commonsense compromise that left both parties at least partly satisfied. Increasingly our management team requested that Lang attend significant NASA-Grumman meetings because of the positive contributions he made to their outcome. Jim Neal, NASA's LM contracts manager, reported to Lang and followed his boss's constructive approach to negotiation and problem resolution.

Early in Thanksgiving week the NASA people invited their Grumman counterparts to share Thanksgiving dinner with them in their homes. Bill Rathke and I were invited to Jim Neal's, where we enjoyed a warm, sunny Texas afternoon in their ranch-style house with Jim, his wife, and their two young girls. We wistfully described our families back home and feasted on turkey with all the fixings. Afterward, we watched football and played on the lawn with the Neal children. This warm gesture by NASA helped create a feeling among us "Grummies" that we were a welcome, valued part of the Apollo program.

As the scope of the job grew, the planning that we had forced ourselves to do became invaluable. Our staffing, office space, and equipment requirements were continually revised upward, and many Grumman meetings were devoted to updating these estimates with Carbee, Whitaker, Stern, and the LM section managers. We also discussed our approach to systems engineering for the LM to comply with program requirements.

Systems engineering was a logical methodology for quantitatively analyzing mission requirements and breaking them into subsets for assignment to systems and subsystems within the spacecraft or supporting ground complex. The systems were designed in response to this array of mission and systems requirements, and its performance was calculated and later verified by tests. A standardized method was used to diagram the physical and functional performance of systems and their interactions and dependencies. Systems performance was documented against requirements and updated with test results. Detailed control of the spacecraft's physical configuration was an essential part of the systems engineering process. The "as designed," "as built," "as tested," and "as flown" configurations were documented down to the smallest part and rigorously controlled throughout the life cycle of a spacecraft.

Stern built on his experience at Arma to make himself an expert in systems engineering, one of a few at Grumman. This alone made him invaluable to me on LM, because NASA had invoked the air force's systems engineering approach wholesale for the Apollo program. Stern brought far more than expertise, however; he had engineering insight that allowed him to quickly deconstruct a complex problem into its essential components and then suggest ways of analyzing possible solutions.

No area of our proposal was more severely criticized by NASA than ground-support equipment. Dick Spinner, our GSE project engineer, took a personal drubbing from NASA for what they considered a poorly conceived, grossly underestimated GSE program. I worked closely with him to reconstruct a more complete version. He needed much reassurance to regain his confidence after NASA's harsh comments. I told him he had done the best any of us could have done, using his experience with support equipment for naval aircraft in the absence of direct knowledge of spacecraft support requirements. He looked at me dubiously, his round face etched with self-doubt, and reluctantly returned to his outspread worksheet.

A major aspect of the negotiations was redefinition of the LM program plan, including schedules and deliverables. NASA declared that our proposal did not contain enough test articles, which were nonflyable mockups or partially functional LMs used for specialized tests of individual or combined systems. Our proposal contained a full-sized mockup, two propulsion test articles (ascent and descent systems), a guidance and navigation simulator, and an antenna test mockup, in addition to the ten flight LMs required by NASA's request for proposals. With NASA's guidance we expanded the propulsion test articles to include both heavyweight "boiler-plate" and flight-weight models, added a full-sized metal electrical/electronic systems test article for functional and electromagnetic interference (EMI) testing in a copper screen room, and a full-sized thermal test article to be man-rated and tested inside the large space thermal vacuum chamber at NASA Houston. Also added to the deliverables list were boiler-plate mass simulations of the LM to be flown as ballast on Saturn booster development tests. A number series was established for LM test articles (LTAs) and test models (TMs) to distinguish them from flight-worthy LMs and mockups (Ms). These additions were costed and scheduled as the month of December 1962 flew by.

NASA also wanted changes to our subcontracting plans. RCA had mounted a major effort in support of Grumman's LM proposal and was rewarded by being proposed for a major share of the work. We offered them to NASA as a "subprime," responsible for complete subsystems that would contain many assemblies and components, some of which RCA would buy rather than make. NASA felt this was unnecessarily complicated and would add cost and retard decision making. They also were concerned that it might dilute Grumman's accountability for the end product.

NASA wanted RCA out of the GSE area, insisting that Grumman take full responsibility. The acceptance checkout equipment would be supplied by GE under direct contract with NASA, and for other systems NASA wanted commonality with command-and-service-module GSE wherever possible. They further demanded that Grumman take back the avionics systems integration role we had proposed delegating to RCA. In NASA's view, systems integration was the technical heart of the project and must be performed by the prime contractor.

None of this went down well with RCA's LM manager, Frank Gardiner. Gardiner was an MIT graduate and had served in the U.S. Navy in the Pacific Fleet and at the Bureau of Aeronautics. He was tall, with wavy salt-and-pepper hair and a tanned, angular face. He had the unusual habit of pushing his handkerchief up his shirtsleeve or coat sleeve, where it remained partly visible. It made me think he was a conjurer and would someday pull a rabbit or something else amazing out of his sleeve. I would not have wanted to play poker with him. Despite his riverboat gambler image, I found that Gardiner was straightforward and a man of his word. During our studies and proposal he was effective in unifying RCA's efforts in support of Grumman across several divisions at different locations, and he was helpful on technical issues, especially in radar and electronics.

NASA also disagreed with our choice of subcontractors for the environmental control system and the fuel cells. Our competitive selections had been very close, but in both cases we had selected another company over the existing supplier for the command and service modules. (We chose Pratt and Whitney for ECS and Hamilton Standard for fuel cells, the reverse of the CSM lineup.) NASA asked us to update the competition for the two top-ranking companies, adding a requirement to maximize commonality with CSM equipment. In a two-week frenzy of activity we obtained revised bids from the subcontractors involved and changed our selections to the CSM suppliers of these systems.

The summary management meetings with the NASA negotiators grew longer as we focused on reaching agreement on the revised program plan, technical approach, and cost and schedule estimates. On 23 December 1962, Dave Lang offered Grumman a definitive LM contract through October 1968 valued at $385 million. Grumman's proposed contract price was $345 million; the negotiations had increased the estimated program cost to NASA by more than 11 percent. Gavin was not ready to agree to NASA's proposed cost number until we had more time to compare the increased work scope that had resulted from the negotiations with the added cost allowances.

Joe Gavin recalled that he was the last to leave Gulfgate Gardens to go home for Christmas. Lang was still pressing him to agree to NASA's offer, saying it was the best Grumman would ever get, but he demurred to consider it further.[3]

We booked what flights were available on Christmas Eve and headed for home. What a wonderful relief to be back with my family again, enjoying the happy holiday season with Joan and our four little boys! Joan had our house beautifully decorated with colorful wreaths, garlands, lights, and ornaments. Her loving welcome and the boys' smiles, hugs, and excitement made me wonder why I had stayed so long in Texas.

On my first day back at work in Bethpage internal battles ensued over the proposed separate LM program "clock number." In the LM proposal Grumman committed to a management innovation: the LM program staff was to be "projectized" by assigning all LM employees to a separate personnel clock number. Joe Gavin was vice president–LM program, a corporate officer dedicated to LM with direct control of the people and resources needed. The proposal described Grumman's LM program organization as a "single-product company" within the corporation. This approach assured NASA that their program would receive equal attention within a historically navy-oriented company. It was modeled after projectized organizations that evolved in the Atlas, Polaris, and other ballistic missile programs at West Coast aircraft companies to satisfy customers' demands for "separate but equal" control over personnel and resources.

The LM clock number was a source of internal friction between the LM program and corporate engineering organizations even before the proposal was submitted. Now the Engineering establishment knew that it was for real, not just a possibility. NASA had pressed us hard for assurances that they and the LM project would have real control over internal priorities and people versus Grumman's entrenched U.S. Navy programs. They made us explain in detail how the projectized LM program organization would work, and how it would draw support from the rest of the company. We felt that we had little room for concessions in our internal negotiations with the Engineering Department.

In meetings with Engineering Department heads and project engineers we explained the reasons for the new organizational approach (our NASA customer demanded no less) and how we thought it could work to the benefit of all. The heads were divided in opinion: some were willing to give it a try, and even saw potential gain in providing special training and equipment for space engineering versus aircraft; others were unalterably opposed because they believed it undermined their authority. The issue was settled in mid-January when Vice President of Engineering Dick Hutton and Chief Technical Engineer Grant Hedrick assembled the Engineering Department management, with Rathke and me present, and directed them to cooperate with the LM project organization as we had defined it. This relieved a great burden from my mind, allowing me and Rathke to concentrate on designing and developing the LM.

Although the clock number issue created much discord within Grumman Engineering, we also felt respect growing for the LM program within the company. Eight aerospace companies had bid for the LM, and by winning this hotly contested prize Grumman had gained greatly in prestige, business base, and future competitiveness. Within Grumman we were no longer perceived as oddball "space cadets" but as a major part of the company's future.

Thanks to a coincidence, word spread quickly at Grumman and on Long Island about the Apollo program. I served on the program committee of the Long Island section of the aerospace professional association, the American Institute of Aeronautics and Astronautics (AIAA). Before Grumman's LM proposal preparation began I made arrangements with NASA for air force major Dick Henry, who was assigned to the Apollo program office in NASA Headquarters in Washington, D.C., to speak on the Apollo program at an AIAA dinner meeting to be held at Grumman-Bethpage in November 1962. When the long-planned day arrived, I was not able to attend because I was in Houston for the negotiations. Due to the timing—two weeks after the announcement of the LM award to Grumman—the response was overwhelming. An average Long Island AIAA dinner meeting drew fifty to sixty attendees; this one had more than five hundred reservation requests.

Grumman decided to accommodate all who wanted to come, seating about three hundred people in our largest speaker-equipped cafeteria, the overflow watching on closed-circuit TV in a second cafeteria. Many of the attendees were from local aerospace suppliers and machine shops, wondering how they could get some LM business from Grumman. For them the Grumman Procurement Department announced that a lunar module supplier day would be held at Grumman as soon as the LM program's needs were better defined. Local press and TV also attended Major Henry's briefing, and the widespread publicity attracted a flood of job seekers to Grumman's employment office, helping us to hire skilled people to staff the LM program.

The LM program became the talk of not only Grumman but also all Long Island. Neighbors and acquaintances went out of their way to congratulate me and to wish Grumman luck on the new venture. I knew we had arrived, however, when one day in mid-January I went onto the Engineering floor to look at the area that was being set aside as interim space for LM. There in a large side corridor I could scarcely believe what I saw: tall stacks of brand-new IBM Selectrics in their cartons, reserved for the LM program! Recalling how I had scrounged to get three of those machines to write the proposal, I was astonished. I did not stop bragging within Grumman about that discovery for weeks.

On 14 January 1963 Gilruth visited Joe Gavin at Bethpage. They resolved the outstanding negotiation items and agreed upon a price and final contract wording. NASA then issued direction to Grumman to proceed with LM development. We were on our way! The contract was not formally signed until early March, at a revised cost figure of $387.9 million.[4]

After the announcement of the LM contract award to Grumman by NASA, we received friendly overtures at all management levels from North American Aviation, which had been chosen a year earlier as the Apollo spacecraft contractor.[5] North American Aviation was responsible for designing and developing the Apollo spacecraft, consisting of the command and service modules, and for integrating the lunar module into the complete Apollo spacecraft stack and ensuring compatibility of the spacecraft with the launch vehicle.[6] North American Aviation designed and built the spacecraft/LM adapter, a truncated conical structural shell that housed the LM atop the Saturn booster at launch and upon which the CSM was mounted. They developed the launch escape system (LES), with its solid rocket and tower of tubular struts, the purpose of which was to snatch the CM with its astronauts safely away from the Saturn if it exploded before or shortly after launch. Together with NASA-Houston, they planned the Little Joe 2 solid-rocket program to flight-test the CM's parachute recovery system and the LES.

North American Aviation was unquestionably the senior partner in the Apollo spacecraft development team, consisting of North American Aviation (CSM and spacecraft integration), Grumman (LM), MIT Instrumentation Laboratory (spacecraft guidance and navigation), and General Electric (Apollo reliability and quality assurance [R&QA] and ACE). NAA was the first aerospace contractor selected for the Apollo spacecraft and was a much larger company than Grumman, and we were very pleased by their welcoming attitude.

Joe Gavin, Saul Ferdman, and I visited NAA's facility at Downey, California, where we were graciously hosted by John Paup, the company's Apollo program director, Norm Ryker, the Engineering manager, and Charles Feltz, the project engineer. They showed us through the aging World War II aircraft factory buildings, which had been used by Consolidated Vultee to build B-24 Liberator bombers during the war and were being spruced up with cleanser and paint for Apollo. The general appearance of these buildings was not unlike Grumman's World War II factories. Upon entering the adjacent Apollo Conference Center, however, the dreary world of wartime leftovers was forgotten in a tasteful display of modern functional elegance.

A large reception area, thickly carpeted and decorated in designer shades of brown, orange, and beige, led into the main Apollo meeting room, a combined conference room and briefing theater designed to hold 150 people. It was carpeted and comfortably furnished and contained the best audio-visual and climate-control equipment then available. Paup advised us to provide such a facility; NASA insisted upon comfort and efficiency in their conference centers, as meetings took up so much of their time.

Back home Gavin and Ferdman had a set of renovation plans drawn up, which, although not the equal of North American Aviation's, were a major improvement over what we had. Nothing happened, though, because top management, usually Titterton, Schwendler, or Towl, was very skeptical of the need for such opulence. The company fathers and their navy customers had grown up and worked in spartan, austere surroundings. They cherished the "hair-shirt" image of frugality and conservatism, which had always suited the navy. How, they asked (not unreasonably), could we expect the navy to be content with second-class facilities at Grumman if they saw a luxurious NASA center right in Bethpage? Did not NASA and the navy both work for the same U.S. government? There also seemed to be an unspoken suspicion that we in the LM program were simply seeking to inflate our own status and egos.

Not until the disastrous LM-1 Customer Acceptance Readiness Review in June 1967 was it obvious that the austere image that played so well with the navy was alienating the NASA management of Apollo. We finally did what our Apollo teammate had recommended more than four years earlier, but by then it was too little and too late.

2

Designing, Building, and Testing

5

Engineering a Miracle

"Congratulations—you certainly look happy."
"It's wonderful. Good luck!"
"You look so excited. I'm really glad for you."

Good wishes poured forth from my relatives and friends during the ten-day Christmas holiday break, further stimulating a dizzy feeling of excitement that I could scarcely contain. I had the aerospace engineer's dream job of the century. Not only would I design and build the first spaceship to land men on another heavenly body, but I was encouraged by NASA to let my imagination run wild and question everything we and they had done in prior studies and the LM proposal. I could start fresh, with a clean sheet of paper, using our past work as a point of departure. Such freedom! Now I could probe some puzzling questions more deeply: With no aerodynamics, why did we propose a smoothly contoured LM? Why not just let its form follow function as it will? How would we fly the landing, and what would we find to land upon? Could two men alone safely launch a rocket ship from the Moon? There were so many fascinating issues to explore—it would challenge every bit of ingenuity and talent I could muster. I longed for the holiday break to end so we could start redesigning the lunar module.

In mid-January 1963 we moved into Plant 25, our new LM Engineering building, built on what had been the softball fields on the north side of Plant 5, the red-brick World War II building that housed Grumman Engineering, Preliminary Design, and the Experimental Shops. With 190,000 square feet on three floors, the empty new building seemed huge. In Grumman style it was plain, even austere, with cinder-block walls painted a light beige, reinforced concrete floors covered with speckled beige-and-black vinyl tiles, and white suspended ceilings with row after row of recessed fluorescent lights. But it was fresh and clean, the polished floors gleaming as bright as our hopes for the audacious project we were beginning. On each floor an almost continuous narrow window ran along each outside wall. The steel-partitioned offices on the periphery had half-height glass panels on their inside walls to allow natural light to penetrate the interior. Half-height partitioned offices were erected across an aisle from the windowed offices in some locations, but mostly the interior floor space was open to accommodate the typical aerospace "bull pen" mass-seating arrangement. The office areas for the group leaders whose engineering groups were scheduled to move in first were in place, along with neat rows of new and shiny beige metal desks and chairs.

Bill Rathke and I moved into a large window office on the second floor in the middle of the south wall facing Plant 5. We had decided that as LM Engineering manager and project engineer we must work together hand-in-glove and felt that sharing an office would enhance communication between us. Our beige metal desks and chairs, the same as those on the floor, faced back-to-back, Bill's looking toward the window and mine facing the Engineering floor. There was a large metal table at the end of the room and a blackboard and corkboard on the adjacent metal partitioned wall. On the other side of the partition was a large conference room primarily for our use. It was filled with new furniture: a metal table at one end in front of a blackboard on the wall, and several rows of not-uncomfortable straight-backed metal chairs with vinyl-upholstered seats and backs.

I had come to know and respect Bill Rathke since he had been assigned to the LM program during the proposal. He was in his mid-forties, short and stocky, with wavy dark brown hair and square, rimless eyeglasses. His quiet demeanor made him seem solemn, unless you noticed his eyes, which were often merry and twinkling. His steady, amiable disposition and dry sense of humor made him easy to work with, and his wealth of aircraft design and project leadership experience provided me with much to learn and emulate.

Bill Rathke came to Grumman in 1943 with a freshly minted degree in mechanical engineering from Iowa State University. He had never been east of Iowa in his life and was greatly relieved to find that Long Island was not the same as New York City. Living accommodations were scarce in Bethpage and other nearby villages, so Rathke found a room in a boardinghouse that catered to Grumman people. A few months later he and three other newly hired engineers rented a former summer house near the water in Huntington Bay Hills.[1]

Rathke loved sports and at Iowa State had hoped to play football. His short stature, flat feet, and poor eyesight made that impossible, but he became the Iowa team's manager, looking after the players' gear and traveling to games. Even the wartime draft would not take him—he was 4F—but they urged him to get a job in the defense industry. He had taken courses in aerodynamics and aircraft design and eagerly interviewed with the airplane company representatives who recruited on campus.

At Grumman he started in structural design, but he showed an ability to design other systems as well and obviously had leadership talent. He soon became project engineer of the W2F antisubmarine aircraft, one of Grumman's earliest successful airborne electronic weapons systems. Rathke had more than once taken a complex aerospace project from preliminary design through development, manufacturing, and into flight—the route we now had to follow with LM. Knowing his impressive credentials, I was delighted that he was modest and unassuming as well.

We soon developed an effective partnership for running the LM Engineering Department, keeping each other fully informed through morning and evening discussions so that we could function interchangeably. We divided the work along whatever lines our backgrounds suggested but tried to avoid becoming specialized.

The one factor I had not considered when I suggested to Bill that we share an office was that he was a heavy smoker of cigars. A nonsmoker myself, I found the smoky atmosphere in our office unbearable because my eyes teared and my eyelids became red and itchy. We got Building Maintenance to install an exhaust fan in the ceiling behind a louver directly above Bill's desk. This solved the problem. I watched contentedly as the clouds of gray smoke swiftly vanished into the louver, leaving the air in the office largely smoke free.

Eagerly I tore into the design work facing us. At the LM contract negotiations NASA had made clear that in selecting Grumman for the job they were not buying the design that we had presented in the proposal. Now that we were under contract they wanted us to redo the preliminary design, using more conservative assumptions on weight and redundancy, and with NASA's extensive advice and approval. Rathke and I got the former LM proposal team together, along with the newcomers to the program, and laid out a three-month program to reexamine and rework the proposal design using new assumptions. The primary new assumptions were an increase in the fully loaded LM target weight from twenty-two thousand to twenty-five thousand pounds and an emphasis on assuring that no single failure in LM would affect crew safety—to be achieved by redundancy or by design simplicity and safety factors. I challenged our people to start afresh and rethink the design, because this time what we designed was what we would *build* and *fly* to the Moon: "This is no longer a proposal drill—this is for real!"

One morning in early February 1963 I received phone call from my wife. I answered with apprehension because she rarely called me at work. She told me in an anxious voice that I should come home right away because my father had been in a serious accident. Outside in the cold, crisp air and bright sunshine, my mind raced ahead to dozens of possibilities, all of them bad. I knew Joan would not tell me to come home immediately unless it was very serious.

One look at Joan's crestfallen face told me the worst. My father was dead from a heart attack while on his way to work on the Long Island Railroad. We

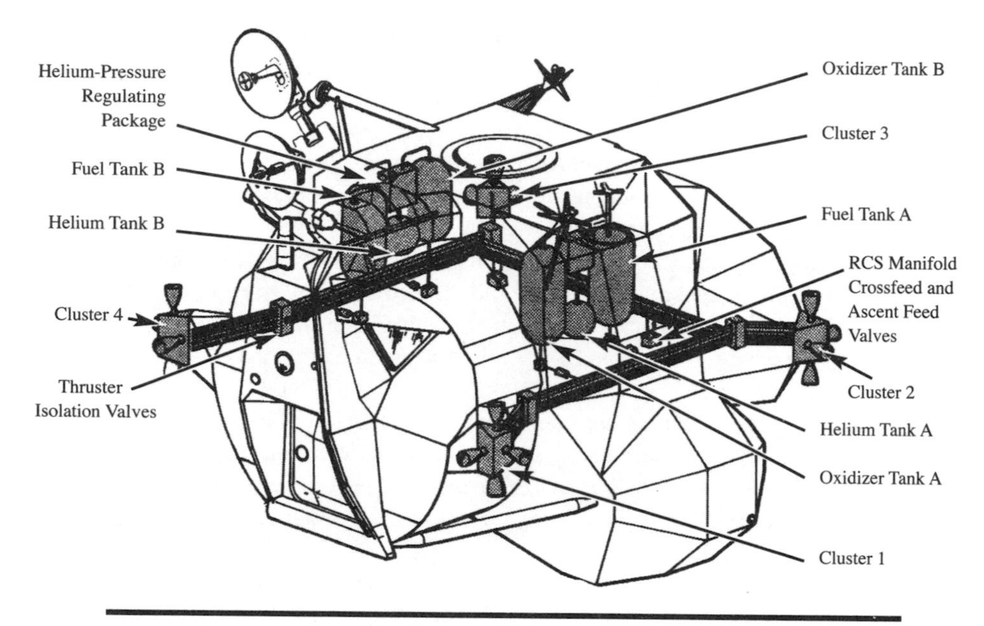

The reaction control system configuration. (Courtesy Northrop/Grumman Corporation)

had to go to the LIRR station in Jamaica, Queens, to identify his body. Shock settled over me, and I sat for a few moments clinging to Joan as both our eyes filled with tears. I pictured my Dad as I had seen him only two days earlier, when we visited my parents at my boyhood home in Merrick. We brought our four children with us: David, eight, Tommy, six, Edward, four, and Christopher, just two and a half years old. It was a lovely visit; both Mom and Dad played with the kids and enjoyed them fully.

At the Jamaica station Joan and I were ushered into a small room in the basement, and there was Dad stretched out on a wooden table, neatly dressed in his business suit, shirt, and tie. He looked like he was taking a catnap, and I hoped that he would open his eyes, get up, and come home with us. He looked healthy, certainly not dead. But he was dead, and I shook with sobs once again as the irreversibility of it all sank in.

The next four days were a blur of grieving, loving, and consolation from relatives and friends. Out-of-town relatives, including my brother and his wife from California and my aunt and cousin and her husband from upstate New York, stayed in the house with my mother, helping her perform the required rituals. My mother, brother, and I received condolences and support from our family, friends, and many people who had known my Dad. At last we laid him to rest with his mother and father in the family plot in Holy Cross Cemetery in Brooklyn and returned home to resume our lives.

My father's sudden, unexpected death was a jolting reminder that life goes on despite the Apollo program. I had been so wrapped up in Grumman that I had lost contact with the "real world." Dad's passing reminded me that I must try to achieve a better balance of home and family life versus work. But my immediate response was just the opposite. In grieving for my Dad, I tried to escape his loss by plunging into my job with greater intensity. The quest to land on the Moon created its own reality, a place where we could isolate ourselves from the problems of everyday life, substituting an intriguing and different set of problems over which we had control.

Defining LM's Basic Configuration

It was an exciting time to be on the LM program at Grumman, and after the immediacy of my mourning had subsided, I bounded into work each day full of enthusiasm. There were many wonderful work sessions that year (1963) in which we used our ingenuity and imagination to make our exotic spacecraft take shape. There were no precedents for what we were doing, so we were neither bound nor guided by convention. NASA allowed us most of the year to firm up the preliminary design and translate it into full-scale mockups.

I vividly recall one Saturday session in the conference room with Ozzie Williams, Bob Grossman, and others from the Reaction Control System Group. We were reexamining the RCS configuration, particularly the geometric placement of the sixteen thrusters on the ascent stage and the degree of redundancy. We spent most of the day at the blackboard, sketching different system configurations and thruster locations and tabulating the features of each arrangement, including weight, redundancy, and controllability after successive thruster failures. We argued engineering logic and practical issues, such as jet contamination effects on adjacent areas of the ascent stage. This latter consideration convinced us to abandon the originally proposed RCS configuration, which had four jets close to the front windows, likely to obscure the windows with rocket-exhaust products.

"Well, where does that leave us?" I challenged. "All the configurations we've looked at have some major drawback. Are there any others that we've missed?" I stood at the blackboard facing the group, all of them deep in concentration.

Bob Grossman jumped up and took the chalk from me, his face glowing. He vigorously erased the blackboard and began to sketch. "Here, how about this?" he said as he continued to sketch. "We'll mount the thrusters at forty-five degrees, keeping them away from the windows. And we'll arrange them as an axial pair and an orthogonal pair. Let's see how that looks in our tables."

Grossman was panting with excitement as he quickly filled in our comparison tables on the blackboard for this configuration. It looked better than any of the others, but we continued discussing its pros and cons for another hour before choosing it as the preferred design arrangement: Four four-jet clusters located at forty-five degrees from the centerline of the crew cabin, with two of the jets aligned parallel to the main rocket engines' thrust axis; the other two jets orthogonal and at ninety degrees to each other. The RCS was arranged in two totally redundant systems, A and B, each having a fuel, oxidizer, and helium tank, eight jet thrusters (two in each four-jet cluster), and fully redundant components. This system could maintain control about all three vehicle axes after any single failure and many selective double failures. We left the office that evening elated at having successfully solved a difficult puzzle and created something novel and valuable.

During this formative phase of the program I realized how fortunate I was to have two very talented assistants who worked effectively with each other and with me. Bob Carbee, LM Subsystems project engineer, was in charge of the design sections, and Arnold Whitaker, LM Systems project engineer, headed the analytical sections. Whitaker led Grumman's proposal team for the navy's TFX missile system and Carbee assisted him. The TFX competition, which included a subsonic attack airplane called the Missileer to carry the advanced missiles, was canceled by the navy in a change of planning direction while the missile proposals were being evaluated. The cancellation occurred as we were preparing the LM proposal, making Carbee and Whitaker available for key engineering positions on the LM program.[2]

Bob Carbee was tall and athletically built, often smoking a straight, sporty pipe, which enhanced his 1950s movie star image. Nevertheless, Carbee was a down-to-earth design engineer with broad interests and curiosity. At Grumman he worked his way up from structural designer to armament group leader, then cockpit group leader, designing and integrating the cockpits with their controls, instrument panels, ejection seats, windshields, and plexiglas canopies. He led a newly formed Weapons Systems Section that cut across older design specialties to analyze, design, and integrate the airborne weapons systems into the aircraft and be responsible for their mission performance. For example, Carbee's section was responsible for the overall accuracy of the A2F-1 Intruder's attack system, the lethality of the Eagle missile system, and the operational effectiveness of the W2F-1 Hawkeye. To integrate these systems across disciplines Carbee became a practitioner of systems engineering, learning this field from the bottom up before the air force's systems engineering manuals even existed. His breadth of systems design experience and exposure to a wide variety of missions and flight vehicles made the LM the next logical step in his career.

Carbee's office reflected his wide-ranging interests. Engineering artifacts were everywhere—cutaways of bomb-release mechanisms, an escape-seat actuator, canopy locking latches, and so forth. Behind each item was a story about the technical issues that shaped its design. On his desk and walls were pictures of his life beyond Grumman. One was of a smooth-faced nineteen year old posed in his leather flight jacket and overseas cap next to a sleek fighter plane. Carbee downed two German planes with his P-51 Mustang during the bloody campaign up the rugged boot of Italy. When I asked him about his wartime service, he modestly remembered being very young and very scared. Other photos showed his lovely wife Mary and two beautiful daughters.

Carbee was a competitive but good-humored sailor racing a seventeen-foot Mobjack weekends in the strong winds of Long Island's Great South Bay. Always among the winners, on Mondays he often told us about his latest adventures, replete with knockdown gusts, hairbreadth recoveries, and a stirring finish.

The section heads who reported to Carbee admired and respected him. He helped them solve technical problems and gave sound advice regarding administrative and personnel matters. Once convinced that they were right, he backed up his people ferociously, tackling me, Rathke, or anyone else in the hierarchy to make sure their viewpoint was considered. He was tenacious in argument, seldom losing his temper unless severely provoked.

Carbee and I led some memorable design meetings in structures and propulsion to review and revise the configuration. The Grumman LM design as proposed just barely fit the constraints imposed by NASA at its estimated weight of twenty-two thousand pounds. When this increased to twenty-five thousand pounds during negotiations, the carefully constructed proposal design house of cards collapsed and had to be reinvented.

Arnold Whitaker was a brilliant systems engineer with an instinctive feel for how to solve problems through rigorous analyses that yielded pragmatic results. He was short, slim, and bald with flashing blue eyes and a sharply pointed nose, and his probing intellect was evident in his persuasive, logical approach to problem solving. Arnold had a disconcerting habit of thinking carefully and deliberately before he answered a question or presented an opinion.

Whitaker and Carbee were the same age, four years older than I. Whitaker served in the navy during World War II but never saw combat. An aeronautical engineer from MIT, he obtained a master's in applied mathematics from Adelphi University. After a brief stint at Chance Vought Aircraft, Whitaker came to Grumman when Vought moved from Connecticut to Texas. He became group leader of Research Control Analysis, developing techniques for analyses of aircraft control systems. Swept-wing supersonic aircraft with hydraulically powered ailerons, elevators, flaps, and rudders demanded increased sophistication in mathematical analysis and computer simulation, and Whitaker was at the forefront of Grumman's efforts in this area. He worked on problems that commanded attention throughout the industry, such as control surface flutter, control reversal at transonic speeds, and powered control system instabilities.

Whitaker was a forceful leader as well as a gifted analyst. He drove studies and analyses hard until they produced answers designers could use in practical systems. He became project engineer for the Eagle missile, an advanced two-stage solid-rocket air-to-air weapon won by Grumman in a navy competition. I remember him analyzing a missile separation dynamics problem at the blackboard with his project group leaders, filling the board with sketches and equations, striding back and forth, head down, thinking, stabbing the air with his cigarette when offering a new approach: an impressive performance that left each participant with an assignment and due date for his part of the puzzle.

The navy canceled the Eagle project just before it was to enter flight testing due to a change in strategic planning direction. It decided to have a compatible series of airborne missiles to perform various missions, launched from a subsonic missile-carrying airplane called the Missileer. Whitaker went from leading a real project to heading Grumman's proposal team for its successor, the TFX(N) missile.

Whitaker played a major role in all aspects of LM Engineering because the analytical sections, including Systems Analysis and Integration, Structural Analysis, Thermal, Weights, Dynamic Analysis, Electronics, and Systems Test, were under his leadership. He and Bob Carbee contributed to many of the most creative sessions during which the LM design took shape and matured.

The LM proposal configuration had cylindrical structure in the ascent and descent stages and a five-legged fixed landing gear. The descent-stage structure housed the descent propulsion system (DPS), the main elements of which were the descent rocket engine, six spherical propellant tanks (three fuel and three oxidizer), three spherical helium pressurant tanks, and associated plumbing. The descent stage also supported the landing gear, provided compartments to carry scientific equipment to be deployed on the Moon, and carried the landing radar. The ascent-stage structure contained the ascent propulsion system, consisting of ascent rocket engine, four spherical propellant tanks (two each fuel and oxidizer), a spherical helium pressurant tank and associated plumbing, and the reaction control system, which had sixteen small rocket thrusters mounted in clusters of four, plus spherical fuel, oxidizer, and pressurant tanks. The ascent stage also housed the crew compartment with its seats, instrument panels, controls, and helicopter bubble windows; the electrical power system (EPS), with its fuel cells, hydrogen and oxygen tanks, busses, and switchgear; the environmental control system, with its water and oxygen tanks, fans, and plumbing; and all the electronic systems, including GNC, communications (with steerable and omnidirectional antennas), instrumentation, and rendezvous radar. The ascent stage had two docking hatches into the crew compartment, one upper (or overhead) and one forward. All this was squeezed into a package so tight that the least change of ground rules or assumptions would cause it to burst.

At twenty-five thousand pounds the increased diameter of the descent propellant tanks meant that a five-legged fixed landing gear that would provide LM with adequate tipover stability would no longer fit within the spacecraft/LM adapter. We had to accept the added complication of a folding, extendable landing gear, but we gained an increase in the allowable diameter of the LM descent-stage structure.

The proposed descent-stage structural arrangement was inefficient for carrying the loads between the LM and the SLA during boost. We needed deep structural beams spanning the full distance across the SLA. This suggested that the basic descent-stage structural arrangement should be a cruciform rather than a cylinder. A cruciform structure in turn demanded four propellant tanks rather than six, and to contain sufficient propellant they must be cylindrical rather than spherical. The cruciform shape also provided a natural place on each end of the cross to mount a rugged tubular framework with which to attach the LM to the SLA and to mount the folding landing gear. The empty areas between the cross were covered diagonally with light structure to serve as storage bays for equipment that could be left on the Moon, or which would be used by the astronauts during lunar exploration. In a relatively short time the redesign of the LM descent stage was basically completed.

The redesign of the ascent stage took much longer because it involved a greater number of complex interrelated factors, including the ascent propulsion system (APS), the crew compartment, and the electronics packaging approach. The optimum solution to these variables was established by performance, reliability, and weight considerations rather than geometry, which had dominated the descent-stage redesign.

One of the most basic variables in the ascent-stage configuration, the number of propellant tanks in the APS, was not seriously challenged until mid-1963 because I and my associates were so ingrained with the principle of symmetry in design. The proposed four-tank APS was conventionally symmetrical. As we saw more clearly how critical the APS was to survival of the crew we reexamined the design to maximize its reliability. We decided to emphasize design simplicity: because the APS itself and most of its major components could not be made redundant due to weight restrictions, APS reliability would be assured by making it as simple as possible. Simplicity, coupled with ample safety factors and extensive ground testing, would be the key.

Our proposed design already had gone a long way toward simplicity. It was helium pressure-fed (no pumps); hypergolic (no igniters); ablatively cooled (no intricate liquid-cooling passages), and operated at a single fixed level of thrust (simple controls). Minimizing the amount of plumbing, components, and joints was also a virtue in APS design because it reduced the chance of leakage, which, due to the APS's small margins on the amount of propellant required to achieve lunar orbit from the Moon's surface, could be disastrous.

Manning Dandridge, LM Propulsion section head, urged me to consider other tank options, including a two-tank version. Dandridge was a tall, ruddy-faced man who possessed an engaging, confident manner and a breezy, informal style of presenting arguments that only hinted at the careful analysis and judgment behind his approach to engineering. A mechanical engineering graduate of Stevens Institute, he designed and tested fuel systems controls and precision components at Aerotec Corporation and a ramjet engine control system for the Navajo missile at Wright Aeronautical Division. Since joining Grumman he had led the propulsion design work on our missile and space proposals, including the successful bids for OAO and the Eagle missile. He spent a year at Aerojet General as Grumman's resident for Eagle rocket-engine development. Outside of work he was a Life Master bridge player, also skillful at cribbage, and he turned his expertise at mathematically based games to good use in solving engineering problems.

Carbee and I held several meetings with our propulsion and structures engineers in which we considered both the four-tank design, either spherical or cylindrical, and the two-tank design, all tanks to be constructed of aluminum or titanium. The two-tank design gave the LM ascent stage an unusual asymmetric appearance because, to maintain the center of gravity on the centerline of the rocket engine, the distance between the fuel and oxidizer tanks had to be in inverse ratio to their weights when fully loaded. (This ratio is the product of the oxidizer/fuel [O/F] density ratio times the O/F ratio required for proper combustion.) The result, Dandridge admitted, looked like LM had the mumps—on one side only.

Appearance notwithstanding, the two-tank design had simplicity on its side and offered a savings in weight, even greater in titanium. I decided to adopt the titanium two-tank configuration, and we made the case for it to NASA.

NASA was cautious about making such a highly visible change and insisted that we do further analyses of off-design conditions. A symmetrical four-tank design remained properly balanced even if the system operated somewhat off the nominal O/F ratio due to differences in tank pressures, line pressure drops, or other factors, but the two-tank design did not. Operation off the nominal O/F caused an unbalanced moment that had to be offset by the RCS thrusters to maintain control of the LM's attitude and trajectory. We analyzed many off-nominal cases to assure ourselves that control could be maintained at all times and that the extra RCS propellant expended would not negate the weight reduction of the two-tank design. Betting that the outcome of these analyses would be positive, I ordered the two-tank design into our M-1 mockup for the review with NASA in September 1963 but did not obtain NASA's approval until December.

Innovative Electronics Packaging

Another exciting area for engineering innovation was electronics packaging. Led by Bob Carbee, instrumentation subsystem engineer Ben Gaylo, GNC subsystem engineer Jack Russell, and RCA Engineering manager Frank Gardiner, we made early decisions that shaped LM's electronics. LM was being designed at a time of pivotal changes in the electronics industry. Circuits constructed of individual transistors were just starting to be replaced by solid-state integrated circuits (ICs) etched on silicon, the beginnings of "circuits on a chip" technology. The new ICs were being developed rapidly for commercial and military use, but they had not accumulated enough operational time to have a solid statistical reliability record. All indications were that the reliability of ICs would be superior to transistor circuits, in addition to saving weight, volume, electrical power, and cooling. I felt we should "bet on the come" and use ICs extensively. After consulting with George Weisinger, our reliability systems engineer, our major subcontractor RCA, and NASA, we established a parts policy that allowed us to use ICs that were just entering the stringent mil-spec parts-qualification test program, as well as those already qualified. If a part subsequently failed the test program, we would have to find a substitute. In practice this rarely happened, and our progressive policy helped keep LM's electronics abreast of the rapidly advancing state-of-the-art until the detailed designs were frozen in 1965.

I asked Ben Gaylo and Jack Russell to develop an Electronics Packaging Specification for LM. This "spec" would establish the shape, geometry, cooling method, connectors, parts mounting, and other standards for the "black boxes" into which most LM electronics would be packaged. With helpful advice from RCA, they developed an advanced packaging design that met my requirements for simplicity, ruggedness, and maintainability. There were two main features of this design. First, wire-wrap mounting of circuits and components onto a mother board. This design, pioneered by the Sippican Company, used a special wire-wrapping tool to wrap together hardened pins from the circuits that protruded beneath the motherboard. The connections remained secure in high-vibration environments, as shown by many convincing tests, yet they could be unwrapped when required for maintenance replacement of the circuits.

The second main feature was circuit cooling by thermal conduction, using copper strips etched into the circuits and motherboards to transfer heat from the heat-producing ICs or transistors to aluminum rails at the midpoint of the sides of the box, upon which the motherboards were mounted. The rails on the boxes in turn were bolted to extruded aluminum mounting rails on the LM's supporting structure through which an ethylene glycol–water coolant mixture was pumped by the environmental control system. The space program pioneered the conduction cooling of electronics. For ground-based and aircraft applications the standard technique was air cooling, using fans or blowers, but in the vacuum of space, air cooling was impractical, and it was undesirable for the low pressure of the LM cabin.

We pushed hard to prepare and issue the electronics packaging specification, which was a constraint to finalizing all our electronics procurements.

When NASA approval was received, we included the spec in our procurement packages and used it in redesigning our electronics installations. In the ascent stage, we designed an electronics rack mounted outside of and behind the pressurized crew compartment. The rack contained parallel rows of cold rails to which the electronics boxes were bolted in vertical columns. Plumbing connections from the ECS provided water-glycol coolant to the cold rails. The area between the equipment rack assembly and the rear wall of the crew compartment was used for mounting water and oxygen tanks and other equipment; the entire area was known as the aft equipment bay.

I enjoyed developing the LM electronic packaging design as we did it, and later I appreciated it even more. The LM electronics that was packaged according to our spec was among the most reliable, trouble-free elements of the program. Other electronics items that were not packaged in this manner, such as the development flight instrumentation (used only on the first three flight LMs) and the cockpit displays inside the cabin were harder to maintain and had higher failure rates.

Crew Systems and the LM Cockpit

The proposal LM had a small, pressurized interior resembling a helicopter's crew station with aircraft seats and instrument panels, but with hand controllers replacing the conventional aircraft stick and rudder pedals. We also had forward and upper hatches, each with docking capability for redundancy. During the LM contract negotiations NASA provided us with preliminary information on the spacesuit and backpack designs and of their estimated requirements for crew equipment stowage and lunar-sample return provisions in the cabin. With these additional requirements the cabin volume we had provided was insufficient. A complete redesign of the LM crew compartment was required.

John Rigsby and Gene Harms, the section head and deputy of Crew Provisions, and Howard Sherman, Human Factors section head, realized that Grumman had much to learn from NASA, McDonnell Douglas, and NAA about crew equipment, human factors, and medical considerations for manned spaceflight. They spent most of the first three months after LM go-ahead with these organizations, learning the accumulated American experience with humans in space. They tried on spacesuits and learned firsthand how difficult seemingly simple tasks became in a stiff, pressurized suit. They discussed the effects of zero gravity and the problems of extravehicular activities (EVA) with astronauts and crew systems engineers from Projects Mercury and Gemini. They saw the Gemini backpacks and working models of the prototype for Apollo. They discussed problems of lighting and equipment stowage, as well as eating, personal hygiene, rest, sleep, and other practical human considerations as affected by the space environment. They learned about different mockup construction techniques and schemes for simulating zero gravity in water tanks or with suspension devices. NASA scientists gave them their latest opinions about the nature of the lunar surface and the problems men would encounter exploring it. With this knowledge and their prior experience with crew systems for airplanes, they returned to Bethpage to redesign the LM.

Bill Rathke, Bob Carbee, and I had several creative sessions with them discussing the conflicting requirements in the design of the crew compartment. I was concerned about the large amount of window-glass area in our proposal design. Glass was very heavy, it was dubious as a structural material, and the large windows allowed too much heat transfer into and out of the cabin. Weight would be a problem because the revised cabin volume was more than twice that provided in the proposal. We agreed that the crew's vision angles to the landing site that we had provided in the proposal must be maintained or increased.

The solution came in a flash of insight as Rigsby, Harms, and Sherman were rehashing the cabin design around a drawing board that showed a three-view (drawing of an area viewed from three directions) of the interior: "What if we get rid of the seats?"

It was a brilliant, paradigm-shattering question that led to a frenzied afternoon of sketching, reasoning, and debating among the three. The next morning they were in my office, clutching several rolled-up vellum drawings and excitedly announcing that they had a new crew compartment design to show. Rathke and Carbee joined me in the conference room as they proudly unrolled the drawings showing two astronauts inside a redesigned LM crew compartment without seats, in the positions they would take to perform every major activity required inside the LM: flying the spacecraft, docking the upper hatch to the CM, doffing and donning spacesuits and backpacks, resting while on the Moon and entering and exiting the LM via the upper and forward hatches. After briefly perusing the sketches we agreed it appeared to be a superior approach.

Without the bulky seats, the usable volume of the LM cabin became much greater. Seats were not required because the LM's mission was relatively brief (initially two days) and the astronauts were at zero gravity while flying or at Moon gravity (one-sixth Earth's) while on the surface. Even LM rocket firings did not exceed one-third G.[3] Some form of restraint would be required while in zero G, and Harms had sketched foot restraints anchored to the deck and a spring-loaded cable and pulley arrangement clipped onto a belt on the astronaut's waist. Handholds and armrests adjacent to the hand controllers provided additional anchor points.

Best of all, the standing position moved the pilot's eyes closer to the window, yielding the same viewing angles with much less glass area. We discussed various approaches to window geometry and agreed to start with a

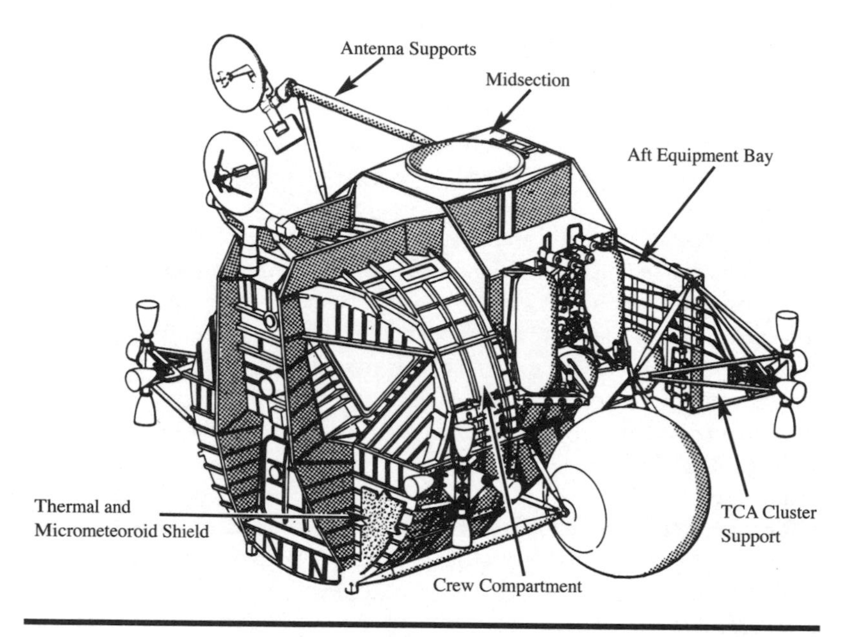

The lunar module's ascent stage. (Courtesy Northrop/Grumman Corporation)

flat window that would provide the same view angles. Will Bischoff and Len Paulsrud joined the discussion and were asked to prepare structural design arrangements for the new crew compartment. Harms and Paulsrud constructed a full-sized foam-board mockup of the crew compartment to work out the combined structural design and crew provisions concepts.

I visited the rudimentary cabin mockup every day as the LM crew compartment that was ultimately built and flown took shape. Computer-aided studies of window geometry led us to tilt the plane of the windows downward and outward, resulting in a small, flat, triangular pair of windows that provided greater vision angles than the original large curved windows. To provide structural rigidity to the flat, pressurized front face of the crew compartment, a pair of deep external beams were added on either side of the forward hatch. A cylindrical protrusion into the cabin behind the flight station was required to accommodate the upper end of the ascent rocket engine, but there was volume behind and on either side of this obstruction for crew equipment, spacesuit and backpack stowage and for rest stations equipped with mesh hammocks. A small rectangular window was added overhead to aid the docking maneuver.

As soon as we shared the new concept with NASA it was enthusiastically received, and a visit from the LM liaison astronaut Donn Eisele further con-

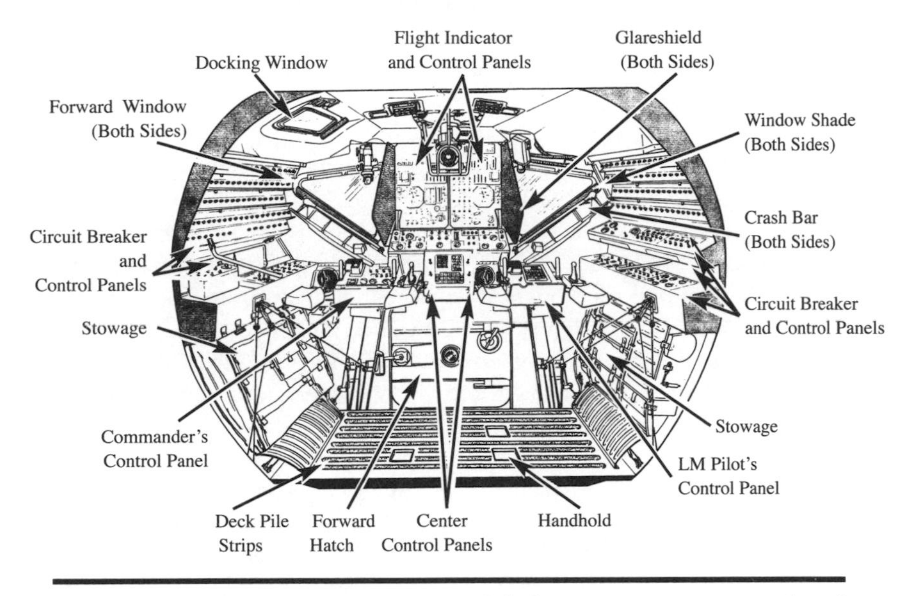

The lunar module's crew compartment and flight station. (Courtesy Northrop/Grumman Corporation)

firmed the acceptability of this approach. (In *Chariots for Apollo,* NASA historians C. G. Brooks, et al. credit NASA crew systems engineers George C. Franklin and Louie G. Richard with originating the idea of eliminating LM's seats. My recollection is that the Grumman team started it, but in any event the NASA and Grumman crew systems teams worked so closely and cooperatively together that the credit is properly shared.) We released the design for the wooden M-1 mockup to be reviewed by NASA in September 1963. An important piece of the LM design puzzle had been put into place.

Mechanical Systems and Explosive Devices

Another design area that required both innovation and careful attention to details was mechanical systems, including the landing gear, docking mechanisms, hatches, equipment stowage compartments and miscellaneous mechanical devices, and explosive devices for stage separation and propulsion-system activation. This was the province of Virgilio "Jiggs" Sturiale, the Mechanical Design section head, and his deputy Marcello "Marcy" Romanelli. This talented pair had many years of experience designing, building, and testing the complex mechanisms required in Grumman's carrier-based aircraft, especially the flight-control mechanisms (for ailerons, flaps, rudders, and elevators), retractable landing gear, wing-fold mechanisms, ejection seats, canopy releases, and tailhook extension and retraction devices. The challenges facing them on the LM were different but no more demanding than the many critical devices to which navy pilots entrusted their lives on every flight of a Grumman aircraft.

Our proposed fixed landing gear contained a major design innovation that, although greatly altered in specifics, carried through into the flight LMs. Concerned about the weight and potential for leakage of fluid into space from conventional hydraulic or pneumatic energy absorbers, I insisted on a dry version. In the proposal we thought it would be some form of molded elastic compound, but as the design evolved into an extendible four-legged landing gear a new energy absorbing material had to be developed to dissipate sufficient energy in a short stroke. Jiggs and Marcy searched for suitable materials and came up with crushable aluminum honeycomb. Hexcell had developed aluminum honeycomb as a lightweight, rigid, high-strength filler material for aircraft control surfaces, particularly the trailing edges, which typically taper down to end in a point. Hexcell was exploring new uses for its material and believed that it could be configured as a highly efficient energy absorber for the LM landing gear. Marcy designed some test struts for which Hexcell fabricated cylindrical slugs of crushable honeycomb. The initial test results were very promising: they indicated that the amount of energy that could be absorbed by a three- to four-inch diameter honeycomb cylinder with a crushable stroke of twenty-four to thirty-six inches was in the range required by the LM's landing conditions. I asked Jiggs to proceed with the landing-gear design using this approach, and we authorized Hexcell to conduct a development program to characterize and optimize crushable honeycomb for energy absorption.

The landing gear Jiggs and Marcy designed was as simple as possible, given the design constraints. It was spring-loaded extensible, released by explosively actuated strap cutters (uplocks), and locked into place by redundant mechanical downlocks on each landing-gear strut assembly. There was no requirement to retract the gear once extended. In the final design the main compression strut on each of the four legs had a stroke of thirty-two inches and the footpad was thirty-seven inches in diameter, a compromise value given the unknowns of the lunar surface that we felt would ensure minimal penetration of the LM into a low load-bearing strength surface.

There were conflicting opinions from the experts regarding the nature of the lunar surface even as LM was being designed. Radar echoes from the Moon hinted at a possible porous surface, which Thomas Gold, respected astrophysicist at Cornell University, interpreted as indicating the possibility that most of the surface was covered with deep, fine dust. Both visual and radar observations, however, showed many areas of boulders and exposed outcroppings of solid rock. This gave us a wide range of potential landing conditions, so we sought a design that would accommodate them all.

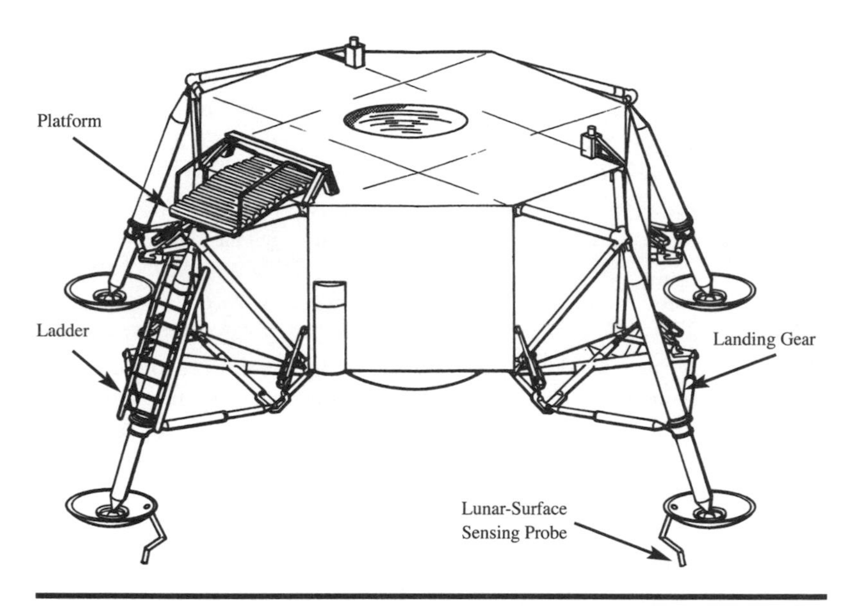

The lunar module's descent stage. (Courtesy Northrop/Grumman Corporation)

To establish the most basic geometry of the landing gear, namely, its tread (distance between opposite footpads) and its height (distance from the surface to the bottom of the descent stage), required a complex systems engineering study. From analyses and ground-based flight simulations, the Flight Controls Group determined the expected tolerances in touchdown velocity that could be expected. These tolerances were expressed statistically as the probability of not exceeding a given value, based upon hundreds of computer and manned simulation runs. After examining the statistics and consulting with NASA and the astronauts, we recommended and NASA accepted the following touchdown parameter limits to be used in LM design: (1) ten feet per second vertical velocity with zero horizontal velocity, (2) seven feet per second vertical velocity with four feet per second horizontal velocity, and (3) vehicle attitude within six degrees of local horizontal.

These touchdown parameters had to be applied to a set of design assumptions about the lunar surface. Here I relied upon NASA, whose experts had been studying what was known about potential landing sites. We arrived at the following lunar surface characteristics for design: (1) six-degree maximum general slope, (2) twenty-four-inch depressions and protuberances, and (3) surface friction coefficients between ice and rock.

Our Structural Analysis Section under Dick Hilderman, working closely with the Structural and Mechanical Design Sections, worked out a multivariable matrix of analyses, starting with different assumed LM spacecraft and landing-gear geometries and covering the full range of touchdown velocities and lunar surface conditions. Many thousands of computer runs were required in this iterative, trial-and-error process to find the bounds of acceptable vehicle and landing-gear geometry for landing. The analyses extended over many months and were still being updated and refined well into 1966. By August 1963, though, we were able to freeze the geometry for the M-1 mockup, and the subsequent dimensional changes to the landing gear did not exceed a few inches.

As we explored the analyses, certain combinations of variables came to be recognized as critical, and we could save time by testing new geometries against these critical conditions first. For example, the critical condition for LM tipover resulted from landing at maximum horizontal velocity (four feet per second) downhill on a six-degree surface slope with the LM attitude pitched downhill at six degrees. The initial surface contacted was ice, but at maximum downhill sliding velocity the footpads hit a solid rock curb. The critical condition for energy absorption and ground clearance occurred when the LM touched down at maximum vertical velocity (ten feet per second) with all four footpads landing in twenty-four-inch-deep craters and twenty-four-inch boulders scattered on level ground underneath the descent stage.

Our designers responded good-naturedly to the outrageous combinations of landing design conditions dreamed up by the structural analysts. They took it as a challenge to overcome seemingly impossible landing requirements, if possible with margin to spare. Occasionally I would intervene, deciding that the likelihood of certain combinations appeared so slight that they should be moderated to avoid unduly penalizing the design. The end result, a lunar module twenty-three feet high and thirty-one feet in diameter across the landing gear, represented a cautious approach to the information available at the time. In hindsight, our landing-gear design proved to be extremely conservative. In the actual missions the astronauts skillfully set the LM down "like a crate of eggs" at touchdown velocities of four feet per second or less. The energy absorbing struts were never stroked more than six inches, and the lunar surface conditions were generally more benign than assumed.

The docking mechanism to connect LM to the CM was an interesting engineering problem. It had to be simple, strong, and absolutely reliable because a jam-up would force the astronauts to the more risky option of transferring from one vehicle to another by EVA. Our proposal design used a double-ended cylindrical docking module that could be mated to the docking rings on the CM and either the upper or the forward hatches on the LM. The idea was to provide docking hatch redundancy on the LM side of the connection.

By mid-1963 North American had begun working on an internal probe-and-drogue docking concept, designed to be assembled and disassembled inside the tunnel between the two spacecraft by the crew.[4] The tunnel in each spacecraft was a thirty-two-inch diameter aluminum alloy cylinder about eighteen inches long located directly above (outside of) the hatches, with machined docking rings on the outer end of each tunnel. The probe mounted in the CM tunnel was driven home into the conical drogue in the LM tunnel and secured by spring-loaded latches. Both the mechanism and the docking maneuver were similar to the air force's probe-and-drogue aerial refueling and very familiar to military pilots. This was a major argument in favor of the probe-and-drogue approach, and Grumman's double-ended docking module was discarded.

After the probe was engaged and latched in the drogue it was retracted, drawing the docking rings on the LM and the CM together and triggering a dozen spring-loaded capture latches, mounted on the CM side, around the circumference of the docking rings to form a rigid structural and pressure tight connection. The drogue was held in place by three mounting lugs in the LM tunnel. It was set into position and later removed by the CM pilot, who also installed the collapsible probe assembly onto mounting lugs in the CM tunnel.

Grumman's Mechanical Design Section worked closely with NASA's astronauts and engineers and with North American to assure that all structural, mechanical, and functional aspects of the hatches, tunnels, docking rings, and mechanisms were coordinated and specified between the two spacecraft. Key physical and functional interfaces between LM and CM were controlled by interface control documents (ICDs), which were prepared and approved by both spacecraft contractors and approved and maintained by NASA. An extensive ground-test program was conducted jointly by Grumman and North American and flight demonstrations of rendezvous and docking took place on Apollo 9 in Earth orbit and Apollo 10 in Lunar orbit. This painstaking attention to detail paid off: the docking system functioned properly on every Apollo mission to the Moon, despite Mike Collins' fears on Apollo 11 that the mechanisms would become jammed in the tunnel and spoil the mission.[5]

Mechanical Design was also responsible for designing a safety critical array of components known as the explosive devices subsystem. They fell into two categories: detonator cartridges, containing explosive charges of high yield, and pressure cartridges, containing propellant charges of relatively low yield. The former were components required to effect ascent/descent-stage separation during launch from the Moon's surface: explosive nuts and bolts that secured the stages together, and the umbilical cutter and circuit interrupter that severed and inerted the interstage umbilical wire and tubing bundle. The landing-gear uplock that held the landing gear in its retracted (stowed) position until fired to deploy the gear was also in this category. In the pressure category were normally closed, explosively opened valves used to release helium from storage tanks into the RCS, ascent, and descent propulsion systems. By containing the helium in its storage tanks and leaving the propellant tanks unpressurized until shortly before these systems were required to function during the mission, the risk of leaking precious propellant or pressurant into space was reduced.

These devices could not be tested before use except for a low-voltage test of the igniter to assure that electrical continuity existed. Therefore their reliability depended upon redundancy in design (dual igniters, explosive charges, nuts and bolts both shattered, dual cutter blades, and so on), rigid process control during manufacture, and statistical sample test firings of components from each production lot. Any test failure caused rejection of the entire lot and an investigation of manufacturing steps to find the cause. Careful attention was paid to grounding and shielding to protect against premature uncommanded firing caused by stray currents or electromagnetic fields. Given the life or death importance of these devices, Sturiale and Romanelli were always among the most nervous Grumman engineers when supporting a flight mission, their worried expressions giving way to grins only after the capture latches clamped home to complete the ascent-stage docking to CM in lunar orbit, the last of the explosive devices having fired upon ascent-stage liftoff. As with their mechanical devices, they achieved a perfect record of explosive device performance on the Apollo missions.

Making LM Reliable

The issue of reliability, so clearly exemplified by the explosive devices subsystem, demanded much of my attention during the formative design period of 1963–64. Foremost in clarifying reliability requirements and expressing them in practical design policies and guidelines were Arnold Whitaker, assistant project engineer-Systems, Erick Stern, manager of Systems Analysis and Integration, and George Wiesinger, manager–LM Reliability Group.

NASA had posed a very broad requirement for reliability: each Apollo mission must provide .999 probability of crew safety (one in one thousand chance of fatality) and .99 probability of mission success (one in one hundred chance of aborting the mission). These overall probabilities had been apportioned by NASA to the individual elements making up the total mission, including the LM. We in turn had apportioned our total unreliability allowance ($= 1 - p$) among each of the LM systems and subsystems, resulting in allowable failure probabilities of one in ten thousand or less for each system.

From a designer's point of view these probabilities were not much help. In practical terms they could not be demonstrated because the allowable failure rates were so low that to prove them would require hundreds or even thousands of repetitive tests. Analyses, however, could be used to show *relative* failure rates of alternative system designs. The absolute value of such analyses was always suspect, but they would indicate the extent to which component redundancy or other system configuration changes would improve overall system reliability.

Reviewing the results of many systems analyses and tradeoff studies, I decided there were a few practical guidelines that we should follow to achieve the highest possible reliability for LM:

1. Specify the highest quality systems and components the current state of the art could achieve.
2. Provide system-level redundancy wherever possible, preferably by dissimilar means.

   Examples of dissimilar redundancy: Lunar-orbit rendezvous, primary method, LM active with rendezvous radar; secondary method, CM active visually sighting LM or its tracking light through telescope; tertiary method, ground-based radar tracking of both LM and CM. Another example: Earth/LM communications, primary method, S-band steerable antenna on LM; secondary method, S-band omni and UHF omni antennae on LM, relay through steerable antenna on CM; tertiary method, LM omni antennae directly to Earth.

   Examples of similar system-level redundancy: reaction control system, fully redundant A and B systems, each capable of maintaining flight control of LM; electrical power system, fully redundant A and B systems, each capable of providing power to all electrical loads, up to half the total ampere-hours of the combined systems. An additional partially redundant bus served essential loads only, and a completely separate redundant system and batteries powered the explosive devices system.
3. Provide component-level redundancy at the highest component level possible.

   Component-level redundancy was provided in most systems, even those for which total system redundancy was impossible due to weight or functionality restrictions. For example, the ascent and descent propulsion systems had redundant valves, regulators, and pressurant lines, even though major components, such as rocket engines and propellant tanks, could not be duplicated. Extensive component-level redundancy existed in the environmental control system, although the cabin pressure shell structure and the spacesuit could not be duplicated.
4. Strive for simplicity and ample design safety margins.

   This guideline became the principal line of defense for systems, such as propulsion, and elements, such as structure and landing gear, that either could not be made redundant or would gain no reliability benefit from redundancy. NASA imposed a program-wide set of structural design safety factors: 1.1 times maximum predicted applied stress before yield of the material, and 1.5 times before failure. These provided adequate margins while recognizing the need for "just good enough" designs to achieve spacecraft weight goals.

5. Test extensively and exhaustively in various environments and stress levels, including stress to failure. Document all failures and investigate until the specific cause is found and design, manufacturing, or operational corrections have been made.

   A particularly useful test was acceptance-vibration testing of systems and components, which tended to disclose both design and manufacturing defects that could be corrected. Joe Gavin led a crusade to refine the design and improve reliability by relentlessly tracking down and correcting the cause of test failures. Gavin proclaimed throughout the program, "There are no random failures; every test failure has a specific cause that must be found and corrected."[6]

We developed these reliability approaches and applied them to LM with NASA's advice and approval at every step. One obvious result was that the number of LM components grew dramatically, accompanied by a major increase in weight. In January 1964 NASA approved increasing the LM control weight to 29,500 pounds (fully loaded, without crew).[7] We agreed to try to achieve a target weight of 25,000 pounds, but the propellant tanks were resized for the control weight. Weight control became more important as the design moved from sketches to drawings to hardware, until in 1965 it became my primary concentration.

An important program-wide NASA decision in mid-1963 greatly simplified the spacecraft's design and lunar mission planning. The competitions for the Apollo spacecraft and for LM both specified in-flight maintenance (IFM) and repair. Built-in test circuits would detect failed replaceable assemblies or components, which would be carried as spares inside the CM and LM crew compartments, and be manually replaced by the flight crews as needed. Although we dutifully complied with this approach in our LM proposal and delegated much of its analysis and implementation to RCA, I never liked it, and once we won the LM contract I tried to change it.

I was convinced that in-flight maintenance would degrade reliability instead of improving it, for many reasons. For one thing, the built-in test circuitry itself was complex and required adding sensors or test ports at critical system locations, which themselves became additional potential failure points. The connectors or mechanical attachments that were required to make the components removable in flight were less reliable than the alternate designs of fixed attachments of hard-mounted components that were only replaced by skilled technicians in a factory clean-room environment. In-flight maintenance made the wiring harness and electrical connectors more susceptible to short circuits and corrosion from humidity and liquid spills in the cabin because protective techniques such as hermetic sealing and connector "potting" (sealing with waterproof hardening putty) probably could not be used. If broadly applied, IFM would require most of the electronic equipment to be located in accessible areas within the crew compartment, increasing its size and internal heat load. Because the spare components and assemblies would have to be stored within the crew compartments, they would be exposed to the humid internal environment. The number of spares of each type would have to be estimated by failure rate analyses that would not be perfect, resulting in payload wasted carrying unused spares that could be more effectively applied to providing redundancy in the basic design. My list of objections was long and, I thought, convincing. Owen Maynard and his NASA LM engineers became as determined as I to eliminate IFM.

Other powerful voices within NASA also began attacking IFM. Houston Flight Operations director Christopher Kraft argued that the crew simply would not have time to repair faulty hardware during LM operations. When George Mueller took over as Manned Space Flight chief in Washington in September 1963, he also had reservations about it. Shortly thereafter IFM was deleted from the entire Apollo spacecraft. Instead the crew would rely on operational displays, the caution and warning system, and ground-based support from the Mission Operations Center in Houston to detect malfunctions. Switchable redundancy would be "wired in," and all electronics inside the cabin would be hermetically sealed or potted to protect against moisture and contaminants.[8] This encouraged us as designers to locate as much electronics as possible outside of the crew compartment, making it smaller and more flexible in accommodating lunar surface mission requirements. I believe this sound NASA decision contributed to Apollo's success.

Project Christmas Present

In the fall of 1963 North American invited Grumman and MIT to join a task force at Downey devoted to establishing an integrated set of Apollo program schedules. The schedules they had prepared in October 1962 had been rendered meaningless by subsequent delays, and as spacecraft integrator they could not properly function without detailed schedule goals. The task force generated the Apollo spacecraft development test plan (ASDTP), the first comprehensive set of subsystem, ground-test, and flight-test schedules linking the CSM, LM, and GNC system with one another and with the Saturn booster. The initial draft of the plan was submitted by the contractors to Houston just before Christmas and was dubbed Project Christmas Present by the task force.

The Grumman contingent at Downey was led by Reynold "Ren" Witte and Theodore "Ted" Moorman. Both were experienced test engineers; Witte's background was in ground testing from the Structural Test Group in Engineering and Moorman's in aircraft flight testing with the Flight Test Department. Both were well-organized and effective leaders. They tapped into the test engineering corporate memory at Bethpage and directed the ten to twenty Grumman engineers on site who were temporarily assigned to support them. Witte and Moorman had led the LM development test negotiations with NASA after contract award and were thoroughly familiar with the intricate interrelations and prerequisites between critical test milestones on the LM subsystems and in LM's flight-development program. In the ASDTP exercise they explored and established the constraints that LM development milestones imposed on the CSM, the GNC system and the Saturn, and vice versa. The ASDTP task force was a most beneficial activity for all the Apollo contractors.

The LM program in Project Christmas Present included ten flight LMs, the first two of which were unmanned, and six LM test articles for ground tests, as follows: LTA-2, launch vibration tests at Huntsville; LTA-10, SLA fit and mass model tests at North American–Tulsa; LTA-1, "house spacecraft" at Bethpage for electronics tests and support of fabrication, assembly, and checkout; LTA-8, thermal vacuum tests at Houston; and LTA-3 and LTA-5, structural and vibration tests at Bethpage. The ground-test program also contained boiler-plate and flight-weight propulsion test rigs for fluids tests in the cold flow facility at Bethpage and hot rocket firings at White Sands. The first unmanned LM flight was scheduled for late 1967, and the ground test articles were planned for use during 1966 and 1967.

Grumman Leads Mission Planning

As our LM designs took shape we encountered questions on LM requirements that could only be answered by a better understanding of how the lunar landing mission would be conducted. The design of the crew compartment, for example, was greatly influenced by the crew's activities on the lunar surface. When would the spacesuits be worn? How many times would they be doffed and donned in the LM cabin, and how much volume would this require? What about the return of lunar samples; size, weight, contents? One set of questions generated others. Determining the required capacity and duty cycle of the EPS called for detailed knowledge of the mission time line; accounting hour-by-hour for what mission activities were in progress and what equipment was turned on. The same was true of thermal analysis, determining the heat loads used to size the ECS. Communications system requirements, duty cycles, and antenna positioning and usage could not be finalized without a detailed mission plan, nor could any of the LM systems. I realized that we must have a mission baseline for our design.

Rathke and I proclaimed the need for a design reference mission (DRM) to give all Apollo contractors a basis for finalizing their system design requirements. We asked Tom Barnes to take our existing lunar-orbit rendezvous studies and expand their mission definition. We also began talking up the idea informally at Apollo program meetings with NASA, NAA, and MIT.

In September 1963, shortly after the approval of a definitive contract with North American Aviation, Joseph F. Shea replaced Charles Frick as Apollo spacecraft program manager in Houston.[9] Joe Shea was brought into NASA's newly formed Office of Manned Space Flight in November 1961 from Space Technology Laboratories by NASA's administrator, James Webb, and D. Brainerd Holmes, associate administrator for Manned Space Flight.[10] Known from the Titan and Minuteman ballistic missile programs as a brilliant aerospace systems engineer, he was made Holmes's deputy and given the challenge of evaluating the competing lunar mission modes. Shea was an articulate man of overpowering intellect, a skilled debater, persuasive in argument and a powerful program leader. He was tall and handsome, with an athletic build and dark Irish good looks—jutting jaw, fair complexion, prominent black eyebrows, and black crew-cut hair. He was dangerously dynamic, so likely to prevail in most arguments that the entire Apollo program depended upon his wisdom and good judgment. Although he could witheringly destroy an opponent's arguments as capably as any trial lawyer, he also had a wonderful sense of humor and enlivened many a meeting with his witty, inventive puns.

When I mentioned to Shea our need for better definition of the lunar mission to pin down LM design requirements, he threw the ball back to me by recommending that Grumman lead a mission study with participation by the other Apollo contractors. Thus was born the Apollo Mission Planning Task Force (AMPTF) in January 1964. With Barnes in charge, the AMPTF set up shop in one of the large Apollo conference rooms in Plant 25 and was joined by team members from NAA, MIT, and NASA-Houston. Tom Barnes was a great team leader. Friendly, constructive, and totally without ego or institutional bias, he inspired confidence and cooperation from the entire task force. Barnes was a talented systems engineer who explored problems relentlessly, asking key "what if" questions that sometimes led to new ways of defining or resolving things. He did it in such an easygoing but provocative manner that others were stimulated to new insights and contributions.

The task force started by defining the basic mission objectives: "Land two astronauts and scientific equipment on the near-Earth-side surface of the Moon and return them safely to Earth." A second objective was to carry at least 250 pounds of scientific equipment to be set up on the Moon and to bring back 100 pounds of lunar soil and rocks.[11] The AMPTF created a detailed description and analysis of all flight mission activities from liftoff to splashdown and recovery—the DRM. They also investigated possible failure modes and contingencies to determine their effect on mission planning and on spacecraft design requirements.

Four months of intensive mission planning and analysis took place with dozens of engineers from NASA and the contractors participating. To make possible precise launch and trajectory calculations using the exact relative positions of Earth, Moon, and spacecraft and figuring the rocket firings necessary to execute mission maneuvers and flight path corrections, it was necessary to choose a specific date for the DRM. The team selected 6 May 1968 for liftoff. Using the same minute-by-minute crew time line planning technique that had been developed on Projects Mercury and Gemini, the AMPTF extended the detailed flight plan to cover three astronauts and two spacecraft (CSM and LM) that for part of the mission functioned independently. Actions required of the crew, the spacecraft, and the ground network to perform the mission were documented in the DRM time line, and trajectory calculations and error analyses were performed to establish system performance and accuracy requirements. The result was the most complete prelaunch mission planning yet attempted for Apollo, providing a good basis for further development of design requirements, operational ground rules, and mission plans.

The DRM clarified the docking requirements for both the CSM and LM. Initial docking and extraction of the LM from the SLA would be carried out by the crew from their couches in the CM. The resulting connection would provide a rigid pressurized tunnel permitting crew access between both spacecraft. Upon return from the Moon, the rendezvous maneuver would be performed by the two-man LM crew, with CM-active backup available, but the docking would be done by the lone crewman in the CM in the same manner as the initial docking. Only the upper LM hatch was required for docking, whereas the forward LM hatch was needed for lunar surface egress.

The DRM became a treasure trove of information for contractor engineers seeking firm requirements to which to design their systems and components. At Grumman we set up a formal process to tabulate the requirements specified or implied by the DRM and compare them with the design specifications for the LM spacecraft and its systems and components; correcting the design specs where necessary to bring them into conformance. This assured that the spacecraft we were designing and subcontracting would be able to perform its overall intended function of lunar landing and return.

The planners looked for failure modes at each step of the mission and sketched out recovery plans where possible. They also determined the accuracy required in critical mission phases. For example, the midcourse trajectory corrections on the way to the Moon had to be accurate within three or four feet per second or else the spacecraft would crash into the surface. Upon reentry the CM must hit the outer edge of the Earth's atmosphere within a flight path angle "window" of two degrees. Too steep an angle would result in a rapid plunge into the atmosphere, burning up the spacecraft like a meteor, while too shallow an approach angle would skip the CM off the top of the atmosphere and send it on an eternal orbit of the Sun.

One major result of the AMPTF contingency planning was the identification of the "LM Lifeboat" mission. While postulating the effect of various CSM failures on the outbound leg of the mission, the planners realized that a number of them could be countered by using the LM as a lifeboat and utilizing its propulsion, guidance and control, life-support, and other systems to return the crew to the vicinity of the Earth's atmosphere for reentry in the CSM. To provide this rescue capability, some of the LM consumables, such as oxygen, water, and electrical power, would have to be increased by 10 to 15 percent above that needed to perform the basic mission. Because LM then existed only on paper, we decided to make the tanks that much larger. At a later date it could be decided whether to actually load the additional consumables into them. Six years after it first appeared in the AMPTF's report, this vital crew rescue mode was dramatically utilized on Apollo 13.

Loose Cannon

As we performed comparative analyses on the LM systems we came to doubt the reliability estimates for the MIT guidance, navigation, and control system that was provided to both the LM and the CM. We arrived at this conclusion while preparing our own estimates of the reliability of the backup abort guidance system, for which Grumman was responsible. This led us to challenge MIT's GNC reliability estimates, with disastrous results for Grumman.

Our GNC experts came to believe, and they convinced me, that MIT's GNC had a factor of one hundred lower reliability than MIT claimed, an opinion based mainly upon Grumman's interpretation of summary mean-time-between-failure (MTBF) data for guidance system components on the Polaris, Titan, and Minuteman ballistic missile programs, as published in a GE report. It was instigated by a former Honeywell reliability engineer working for Grumman who may have had a hidden agenda. (MIT had beaten Honeywell in the competition for the Apollo GNC and said engineer was on the Honeywell proposal team.) After Grumman published its conclusions in a report to the Apollo program office, MIT and NASA hit the roof.

Joe Shea convened a meeting of all interested parties in early January 1964 to find the truth and punish the guilty. We gathered in the well-appointed Apollo program conference room in Houston. About thirty NASA, MIT, air force, and Bellcom GNC experts attended, led by Jim Elms, representing Bob Gilruth, MSC director, and Joe Shea, Apollo spacecraft program manager. Gavin, Rathke, Whitaker, and I sat opposite this stone-faced group. Shea warned everyone that the meeting was being tape-recorded.

Elms welcomed everyone and said the meeting would be a technical discussion to resolve questions concerning Apollo guidance system reliability. He turned the meeting over to Shea. Shea glowered under his black eyebrows and hunched forward toward his microphone, his hands knotted tightly together on the desk, his gold MIT ring visible:

> Gentlemen, we have a serious problem. The problem is that Grumman believes the MIT guidance system is two orders of magnitude inferior to other available

> systems, and that the Apollo program is being jeopardized by this choice. The issue centers upon the evaluation of the data used in drawing this conclusion and establishing its validity.
>
> I intend to force a black-and-white conclusion as a result of this meeting. Either there is or there is not a significant basic difference in the inherent reliability of the MIT system and other comparative data. Someone, Grumman or MIT, will have to leave this meeting admitting he was wrong—mea culpa, mea maxima culpa.

Shea beat his breast for emphasis. He explained the points of interest in reliability data assessment, lecturing us like schoolboys. The sources, accuracy and time frame of the data, the relevance of failures included, and the type of program all must be interpreted to select data that is comparable to the Apollo system.

Whitaker took the stand to present Grumman's case.

He said that our evaluation of the inertial guidance system reliability data that we had been able to obtain predicted 894 failures per million hours, versus 10 failures per million hours estimated by MIT. This puzzled us and caused us to consider other design approaches for the LM abort guidance system, such as strapped-down inertial components instead of a gimbaled platform. In discussions with MIT and with guidance system manufacturers we had not been able to resolve this discrepancy. If Grumman had misinterpreted the data or overlooked other relevant data sources, we were open to correction.

Whitaker reviewed the Polaris data that Grumman used. Most of it had come from Minneapolis Honeywell, which manufactured both a MIG gyroscope (gyro) that they designed and the MIT-designed IRIG gyro. Honeywell's numbers showed a much lower failure rate for the MIG gyro.

Next Whitaker turned to Titan program data. Grumman had only found a small amount of Titan data, and it showed high failure rates. Because the Titan GNC design was similar to MIT's Apollo system, we used this data in calculating our failure rate.

"Didn't you think there must be a lot more Titan data available than that?" bristled Shea, who led the development of the Titan II guidance system at General Motor's AC Electronics Division. "You're implying that all the Titan data you couldn't find also showed high failure rates. Well you're wrong, as you'll soon find out."

Things got worse. When Whitaker showed our use of Minuteman data, the air force said that none of it was applicable because it all related to the percentage of the missile force available on thirty seconds warning—it did not relate to predicting design reliability. Whitaker sat down dejectedly.

Dave Hoag gave MIT's rebuttal. Starting with Grumman's number, 894 failures per million hours, he patiently showed how that number dropped as each incorrectly included or omitted set of data was corrected. Grumman's data was based upon 130 reported failures during 158,000 hours of operation, during the time period covered by the GE report. MIT had more complete results from GE, listing total failures and correcting errors in the report, and these showed 118 failures during 380,000 hours. Eliminating failures of rebuilt units and of designs subsequently scrapped dropped the relevant failures to 60, yielding a failure rate of 196 per million hours. From there on, differences between the design details of the Polaris, Titan, and Minuteman systems from Apollo had to be accounted for. Step-by-step Hoag explained the rationale for these corrections and showed the resultant drop in the failure rates. After two hours of logical explanation he had walked the MIT failure rate down to 20 failures per million hours. He admitted that their published estimate of ten per million was insupportable—attributed to program manager's optimism.

After some general discussion of the responsibility for preparing and using reliability estimates, Shea drew the meeting to a close. "Well," he said, "I think this meeting has accomplished its purpose. Joe Gavin, is there any doubt in your mind that your people were wrong?"

"None whatsoever. I'm sorry we gave everyone so much trouble." Gavin was ashen-faced.

MIT had blown Grumman's analysis out the water, showing that we had not dug deeply enough to properly understand and interpret the GE reliability data. We left the meeting with our heads hung low, figuratively beating our breasts.[12]

This incident had a lasting negative effect on Grumman's reputation on the Apollo program, and for a time it did not do my own personal reputation any good, either. Shea and others saw Grumman as a loose cannon meddling outside its jurisdiction—definitely not team players. They thought we could not be trusted and were particularly upset by what they saw as Grumman's "holier-than-thou" attitude, ready to point an accusatory finger at the alleged transgressions of others. For years afterward we endured strained relations with MIT; not until we were jointly supporting flight missions with them did feelings improve. This incident reinforced NASA's growing belief that Grumman required close oversight and constant detailed direction.

Combining Innovation with Discipline

As 1963 progressed Rathke and I realized that our initial headlong approach to LM design must become more disciplined and systematic. At first I was enthralled by the thrilling feeling that we were inventing something totally new and historic. "No one knows what LM should look like; no one's ever done this before," I told my staff at more than one early project meeting. We were simultaneously awed by the challenge and stimulated with the possibilities for achievement. But the ever-growing complexity of the design and the need to schedule and control activities and determine personnel skill requirements and assignments demanded a more rigorous technique of engineering leadership.

First we established a regular series of LM Engineering meetings. The weekly project engineering meeting was chaired by Bill Rathke, or me in his absence, and was attended by the LM System and Subsystem managers, the assistant project engineers for Systems and Subsystems (Whitaker and Carbee), and direct staff reports, about thirty people in all. At this meeting we reviewed items of general interest, including the previous week's accomplishments and plans for the coming week. We discussed those problems or activities that required major emphasis, making key personnel assignments as needed. NASA's latest directives, trends, and planned activities were also reviewed. The meeting was planned not to exceed one hour and a summary weekly report was published, emphasizing items discussed at the meeting but also including more routine reports from each Engineering group.

I chaired a daily series of technical staff meetings (TSMs) in my conference room, beginning at eight o'clock each morning. These meetings, which generally involved eight to twelve engineers from specific System or Subsystem groups, were progress review, problem-solving, and management-direction meetings, often focused on a single problem or issue. Carbee and Whitaker attended most of these daily meetings, and Rathke was there if the subject warranted. The appropriate system or subsystem managers and their deputies participated on their scheduled days, and Erick Stern or others from Systems Analysis and Integration attended most meetings. The TSM schedule covered every LM Engineering group at least once every two weeks.

These TSMs were scheduled for one hour but took longer if I thought the discussions were productive. I tried to keep the first two hours of the day, until ten o'clock, clear for these meetings. Conducting the TSMs every weekday was arduous and time-consuming but well worth the trouble. Many of the successful approaches to LM design were thrashed out at these working sessions. I tried to publish an agenda with the subjects to be covered at each TSM one cycle in advance, but these were often changed or modified by the press of daily events. The weekly program meeting, which Joe Gavin held at ten o'clock on Tuesdays, did not conflict with the TSMs, but NASA reviews, visits, and meetings frequently did, making TSM postponements and rescheduling commonplace. I clung doggedly to the TSM discipline throughout the four years that I served as LM project engineer and subsequently Engineering director, and I believe that this simple technique allowed me to provide "hands-on" guidance and leadership, and to pull the many diverse threads of the LM design and development together into a coherent whole. I viewed the LM Engineering staff as a marvelous symphony orchestra in which each section (system or subsystem discipline) must excel with its instrument (knowledge and skill), and where each would have a turn to solo (solve a problem or achieve a key milestone) upon which the success of the entire ensemble would depend. When their moments in the spotlight came, no LM Engineering group was ever found wanting; they all gave "star quality" performances.

Another vital technique for LM Engineering was the configuration control process. NASA had adopted a strict, rigorous specification and handbook for configuration control, which had originally been developed in the Department of Defense for the air force and navy ballistic missile programs. Although these requirements were new to Grumman, we were convinced they were necessary to achieve the quality and reliability levels demanded by the Apollo program, and we sent a number of key supervisors to be trained by NASA in the new professional specialty of configuration control. We formed Configuration Control to implement and interpret these requirements, but all Engineering groups were affected by this area.

Besides documenting and controlling the detailed elements of the design, Configuration Control formalized the systems engineering process. All system performance and functional requirements, starting with the very top level description of the Apollo program's mission and objectives, were progressively broken down (partitioned) into more detailed design and performance specifications at lower and lower levels of the system. Finally this process reached the level of procurement and design specifications and drawings, from which parts, components, and assemblies were procured and built. This also was a new discipline to us that we implemented vigorously.

Key to this process and its documentation was a defined set of engineering drawing levels. For the LM spacecraft, level 1 was a single large drawing that showed, as interconnected boxes with annotated inputs and outputs, all the systems and subsystems that made up the LM and their predominant interactions. The level 2 functional diagrams took each system or subsystem and indicated their major elements and components and the inputs and outputs from each on a single large drawing.[13] Level 3 took the system or subsystem down to each functional element with detailed inputs and outputs and reference to performance specifications. Levels 4, 5, and 6 were the more familiar engineering design drawings used to manufacture and assemble the parts at progressively lower levels of detail. These were referenced upward to the higher level drawings.

The Level 2 functional diagrams were especially useful during the first two years of the LM program, when the design was maturing and we were making major choices in design approaches and functional redundancy. These diagrams were a principal tool in assessing the relative analytically determined reliability of alternative system configurations, and they served to document the selected result. The level 2 diagrams were prepared by Erick Stern and his Systems Analysis and Integration Group. As project engineer I had the final approval signature, which I applied after a review meeting with Carbee, Whitaker, Stern, and the appropriate system or subsystem managers.

One day in late November 1963 I was holding such a review on the rendezvous radar subsystem with a group that included Grumman's LM radar experts and Frank Gardiner of RCA. We pored over the drawing and discussed it for about an hour and a half, at which point I considered it ready to approve once a few minor changes were made. I then joined Joe Gavin, Bob Mullaney, Bill Rathke, and others at a meeting in Joe Gavin's conference room where Gavin and Mullaney were relaying to us the results of a NASA Apollo management meeting they had attended in Houston the previous day. Suddenly the door burst open and a stricken-faced man announced, "The president's been shot! In Texas."

In the shocked silence that followed, Joe Gavin, breaking his usual reserve, said heatedly, "That's not true! I just heard the news on the way over here, and it's not true."

Then Joe hurriedly retreated to his office to make calls, as did several others, leaving the rest of us stunned and anxious. One by one they filtered back into the meeting, their doleful features confirming the worst. Joe returned and dejectedly said he was wrong, the president had indeed been shot and was dead. Gloom took over. The meeting was adjourned as no one any longer had the heart for it.

When I reached my office, Erick was waiting with the corrected rendezvous radar level 2 diagram. One look told me that he had heard the bad news, but he said that because President Kennedy himself set the schedule deadline for this project, he would want us to keep pushing ahead.

Alone together in the office we quietly verified that the agreed changes had been made, and I signed the diagram and affixed the fateful date: 22 November 1963. When I finished the Engineering floor was almost empty; the news had come late in the afternoon and most people's reaction was to go home, seeking the solace of private grief. The martyrdom of the young president who had started the Apollo program was a powerful motivator to everyone to overcome all obstacles and reach the goal he had established for us.

Other LM Design Changes

There were other significant changes in the LM design in late 1964 and into 1965. The circular forward hatch with cylindrical tunnel and docking ring was changed to a larger rectangular hatch without a tunnel or docking provisions. This change was made after the M-5 full-scale metal mockup review in October 1964 demonstrated that a spacesuited astronaut wearing his backpack could not get through the circular hatch without undue effort. Abandoning docking redundancy on the LM was rationalized because the command module only had one docking hatch, which was just as likely to fail as a lunar module hatch, in which case EVA would be required anyway. The rectangular shape and absence of the tunnel made forward hatch ingress and egress much easier and allowed us to provide a small platform at the top of the ladder. This made getting to and from the surface safer and easier.

The LM control weight was increased again in November 1964 to thirty-two thousand pounds at Earth launch (without crew), and the propellant tanks were resized accordingly. This increase was made possible by NASA's further refinement of the cislunar (Earth/Moon) trajectories and a decrease in the assumed propellant allowance to the CSM for rendezvous. The LM descent-stage hover time was also reduced by one minute, to ninety seconds. This LM weight increase was the limit that could be squeezed out of the Saturn 5. The LM's weight growth continued unabated each month, casting the specter of a difficult, costly, and schedule-busting redesign effort before us. Within a few months it would become my overwhelming preoccupation.

Additionally, in February 1965, after months of design studies, NASA approved the switch of the LM electrical power system from fuel cells to batteries. This was a change NASA had championed, as they became increasingly concerned about the growing complexity of the hydrogen-oxygen fuel-cell systems under development for both the LM and the command and service modules. The original LM proposal used fuel cells, and after contract go-ahead Grumman held a competition that was won by Pratt and Whitney, which was also building the fuel-cell system for the CSM. The LM fuel-cell system was under development at Pratt and Whitney for two years when it was canceled in favor of the "all battery" EPS.

Fuel cells generate electricity by reverse electrolysis, combining gaseous hydrogen and oxygen in a catalytic reactor containing fine-mesh nickel screens to produce electric power with water as a byproduct. Such a system was attractive for Apollo because it was relatively lightweight, producing more kilowatt-hours per pound than comparable battery systems. The longer the mission duration and higher the total kilowatt-hours of electrical power required, the greater was the fuel cell's weight advantage over batteries. The waste water from the fuel cells could be used for cooling and possibly as drinking water, an additional advantage.

The main disadvantage of fuel cells was their relative complexity. They required tanks for hydrogen and oxygen, the reactor chambers (cells) and associated plumbing, valves, pressure regulators, controls, and instrumentation. Batteries, on the other hand, were self-contained "black boxes" that required only monitoring instrumentation and recharge circuitry to be integrated into the EPS.

Owen Maynard, NASA's LM Engineering manager, was skeptical of the LM's need for fuel cells from the beginning of the LM program. He asked me why we had proposed fuel cells rather than batteries in one of our first meetings during the contract negotiations in Houston. I justified our choice based on the weight advantage—more than 450 pounds—offered by fuel cells. A year and a half later, as the fuel-cell assembly (FCA) weight began to increase as the design progressed, Maynard asked me to revisit the FCA-versus-batteries tradeoff study. While FCA weight was increasing, advances in non-rechargeable battery technology showed improved kilowatt-hours per pound and reduced weight.

Grumman and NASA jointly worked the tradeoff studies for more than six months, looking for ways to make the simpler battery system more competitive with respect to weight. We were able to reduce the total LM power requirements for the mission by minimizing the LM's active equipment during the translunar coast phase and eliminating the need for the LM EPS to recharge the batteries in the crew's backpacks. We simplified the system further by determining that the batteries in the descent stage could be passively cooled and by repackaging from five descent-stage batteries to four. Comparative reliability analyses confirmed our guess that batteries provided a substantial relative improvement in reliability.

Joe Shea became personally interested in these studies when the fuel cells under development for the CSM began to encounter problems during early testing. He encouraged Maynard to work intensively with Grumman to provide justification for switching LM to batteries. (There was never any possibility of using batteries for the CSM because its longer active-mission duration, ten days versus two days for LM, made the weight penalty insurmountable.) By mid-February we had a battery system design that was only two hundred pounds heavier than the comparable fuel cells, which we considered an acceptable price for the improved reliability and operational simplicity of batteries. Maynard and I jointly recommended the change and Shea approved it after examining the study results and the forecast impacts on the LM's schedule.

The switch from fuel cells to batteries was the last major change in the LM's design, with the exception of the block changes that were made later to extend the lunar surface stay time and add the lunar roving vehicle (LRV) and more scientific equipment capacity to the last four LMs. From early 1965 the LM Engineering task transitioned from the heady, creative challenge of designing a manned spacecraft that functioned only in space or on the Moon, to the dogged, grinding work of finishing all the minute details of the design and proving by test and analyses that it could do the job. My attention as LM chief engineer focused on a major triad: getting the drawings out, bringing the LM weight under control, and resolving critical technical problems. The fun part of LM Engineering was over. What remained was the practical design, development, and testing that would determine whether LM would perform its bold mission or end as a failed monument to mankind's intellectual arrogance. We had succeeded in gradually making the impossible look probable and in dividing production of LM into basic jobs that ordinary people could perform. I was confident that we were capable of the monumental tasks that lay ahead.

6

Mockups

I enjoyed building model airplanes as a teenager, even though I was not skilled at it. I particularly liked working on the finishing touch: covering the balsa-wood framework with a paper skin and using dope or paint to shrink it tightly over the structure. I often worked with a classmate who excelled at making models, and he helped me build some fairly good looking ones. Most of my models were detailed replicas of real airplanes, which I chose for their graceful appearance. They were flyable, rubber-band-powered models, and I made sure they could all fly adequately, although my main interest was in building rather than flying them. Most elaborate of all my models was a Waco low-wing monoplane. It had an enclosed two-place cockpit with sliding cellophane glass cover, a radial engine, and streamlined wheel pants covering the fixed landing gear. Painted dark blue and white, it was probably my best-looking model airplane.

Now to my great delight I realized that I would be building models again—this time full-sized mockups of the LM. We needed the mockups in order to fit check the LM's complex shapes and assemblies and to determine whether astronauts and ground crews would be able to perform all required functions during the mission and during prelaunch preparations and maintenance. At contract negotiations in Houston we had agreed upon a long list of full and partial LM mockups. In the ensuing months we shortened the list to concentrate on three mockups during the first year of the LM program: M-1, a wooden mockup of the ascent stage and crew compartment; TM-1, a wooden model of the complete LM; and M-5, a detailed metal model of the entire LM.

M-1

As our ideas took shape for the standup crew position and cylindrical flat-faced crew compartment with canted triangular windows, we checked out their feasibility in a simple wood-and-foam-board mockup of the forward interior portion of the cockpit. We converted this mockup into drawings and sketches from which the more complete M-1 could be made, adding the tanks, rocket engine bell, electronic equipment bay, antennas, and other external features of the LM ascent stage. Throughout spring and early summer 1963 the engineering design groups added to this mockup design definition. A formal review of M-1 was scheduled with NASA for mid-September.

A wave of anticipation swept over me when I saw the NASA review team list: in addition to Gilruth, Faget, Maynard, Rector, and others with whom I was acquainted or had regular dealings, it included the Mercury astronauts and Walter Williams, director of Manned Spaceflight Operations. The Mercury Seven—Scott Carpenter, Gordon Cooper, John Glenn, Gus Grissom, Wally Schirra, Alan Shepard, and Deke Slayton—were world-famous spaceflight pioneers. They would be right here working with us in Bethpage, just like professional colleagues! I looked forward to meeting them and profiting from their precious experience in space.

The M-1 was displayed in Grumman's mockup room in Plant 5, where the navy's mockup reviews were also held. Plant 5 was a large, roughly square, high-bay area with a polished black-asphalt tile floor and white cinder-block walls. Large steel trusses, visible under the ceiling, supported the roof, and rows of fluorescent light fixtures were suspended from them. A traveling overhead crane was available to move large mockups throughout the room. The ceiling was three stories high. At the second- and third-story level on one wall were exposed balconies and catwalks leading into interior laboratories and the Preliminary Design center, protected by waist-high chainlink guardrails. This area was known as the "hanging gardens." From the balconies one could look down on the whole panorama of a mockup review in progress. Spotlights were mounted on the roof trusses and the balconies when needed for extra illumination of the mockups.

The M-1 did not need any special lighting, but we mounted the ascent-stage mockup on a low platform and added steps to the platform for easy access. Several rows of collapsible metal chairs with armrests were arranged facing a metal conference table and lectern in front of the mockup. A small speaker and microphone on the lectern provided voice amplification in the cavernous hall.

On 16 September 1963 the mockup room was full of NASA and Grumman officials and participants, sending a busy hum of conversation up to the exposed rafters. George Titterton and Joe Gavin welcomed our visitors to Grumman's first formal review by NASA on the Apollo program, and Bob Gilruth responded that they were glad to be here and pleased to have Grumman on the Apollo team. Several more conference tables and chairs had been set up to accommodate the large NASA delegation. The Mercury astronauts and Walter Williams sat front and center next to Gilruth, Maynard, and Rector. The astronauts were kept busy signing autographs for awed Grummanites until the meeting was called to order.

For the next two days the astronauts and engineers explored every aspect of M-1. Our Crew Systems Section leaders, John Rigsby and Gene Harms, and Human Factors Section head, Howard Sherman, together with their NASA counterparts, George Franklin and others from the Houston Crew Systems Group, guided the astronauts through various drills inside the LM cabin. They tried getting in and out through the front hatch wearing a simulated backpack (no good, too tight), checked out the hand controllers and visibility out the windows from the pilot and copilot stations, evaluated the displays, controls, and equipment stowage within the cabin, and tried and discussed alternative schemes for providing restraint while piloting LM and while resting and sleeping. The astronauts liked the standup crew stations and the wide field of view provided by the windows. They asked whether the forward landing gear footpad would be visible out the window—it would. We drew a chalk circle on the floor with the footpad's proper size and position so they could see it from inside M-1.

For restraint at the pilots' stations Rigsby and Harms demonstrated a spring-tensioned cable-and-pulley arrangement, anchored to the cabin floor and clipped onto either side of a belt each astronaut wore. They provided slipper-shaped foot restraints into which the astronauts' boots would fit while they stood in the pilot station. With minor adjustments and improvements this met with crew approval. There was less agreement on the rest and sleep provisions. M-1 showed the crew leaning against the rear bulkhead behind the cylindrical table-high cover over the ascent rocket engine, restrained for sleep by an open-mesh hammock. This met with grudging tentative approval, accompanied by a plea for us to do better if possible.

As the review progressed my colleagues and I sensed a difference in attitude toward Grumman among some of our celebrity visitors. Some of them, notably Bob Gilruth, John Glenn, and Wally Schirra, were very welcoming and supportive to us as the newest members of the manned spaceflight team. They seemed eager to share with us experiences they considered possibly useful in designing the LM, and they did it straightforwardly, treating us as professional colleagues and peers. Others, particularly Walter Williams and Alan Shepard, made it clear that they thought the Grumman people were space "greenhorns" who had much to learn and were not in the same league as themselves. This attitude was conveyed indirectly, as I overheard snatches of their conversations with one another and watched whispering, elbow digging, and sniggering when Grumman people were presenting their views. Some of them were condescending; others were even rude. Maynard and Rector noticed some of the byplay and counseled us to ignore it. They told stories about some of the Mercury astronauts to assure us that it was nothing personal against us or Grumman—they were like that with everyone.

On 18 September the Mockup Review Board met, with Maynard as chairman, Rector, Chris Kraft, and Deke Slayton for NASA, and Carbee and me for Grumman. There were only a few dozen chits and they were readily disposed of—most called for minor changes or further study of known problem areas. One thing the astronauts insisted upon was identical displays at both crew positions, including the eight-ball flight attitude indicator. M-1 showed some displays shared between pilot and copilot, but this was clearly unacceptable to the veteran astronauts and we acquiesced to their wishes.

The board approved M-1 with the agreed-upon changes. I was relieved that our basic crew compartment design approach was acceptable to the astronauts and that we could proceed with it. We had jumped over our first major Apollo program hurdle, clearing the bar with many inches to spare.

At home Joan was nearing the end of her fifth pregnancy, and that night she was very restless and uncomfortable. In the morning she asked me not to go right into work because she felt something was about to happen. By midmorning we were at Huntington Hospital, and not long afterward she delivered a beautiful nine-pound baby girl, whom we named Jennifer. The baby was delivered by the doctor under the old rules, which meant I was not allowed to participate in any way. I learned the good news from a nurse as I waited anxiously in the hospital lobby. They took the baby away from Joan right after birth (why did we let them get away with such things back then?), but she soon demanded to see her, and the nurse relented and brought in our newest treasure. I saw the little darling through the glass window in the nursery.

Having just completed the M-1 review, I took a couple of days off from work and proudly drove Joan and Jennifer home in brilliant sunshine to the greetings and (germ-laden) kisses of our four boys. We had so much to be thankful for. I had it all: a talented, beautiful, loving wife; five marvelous growing children; and a fulfilling, exciting career. What more could one ask for?

TM-1

Our next mockup was a full-sized wooden model of the complete LM, descent and ascent stages, containing as much engineering detail as we could get into it before the review in March 1964. Its focus was on the crew compartment—especially the support and restraint, displays and controls, equipment stowage, and lighting—and on egress to the lunar surface. We were able to include realistic mockups of some equipment, such as the ascent and descent engines, environmental control system components, and radar and communications antennas. Working models of the hand controllers with which the astronauts would fly the LM were provided at both pilot stations, allowing the crew to experience the tactile feel of the controls as they stood in flight position hooked into the support and restraint devices.

All the LM design configuration decisions that had been made since M-1 were rushed into TM-1, often by using sketches or engineers' "arm-waving" in the mockup shop, so that by the time the mockup review convened on 24 March they saw an up-to-date LM as it was then defined. Only Deke Slayton of the Mercury astronauts attended this review because the others were busy with their crew assignments in Project Gemini, which was at a peak of flight activity. Several astronauts from the second group, selected in April 1962, participated, with Ed White and Charles "Pete" Conrad designated to give LM special attention.

Grumman had learned from the M-1 review that the astronauts had to be treated as a special breed. We needed to communicate with them through another pilot; no Earth-bound engineer or manager, no matter how capable, would command their full respect. Jack Stephenson, our LM consulting pilot, was asked to act as liaison to the astronauts, to make that his priority while continuing the engineering consultation and simulator flying and development that had been his primary activities. Stephenson was an experienced navy pilot and honors graduate of the Air Force Experimental Test Pilots School. With more than forty-four hundred hours of flying time in twenty-seven types of military aircraft, he was in the same experience bracket as the astronauts. At Grumman he was the project pilot on the A2F-1 Intruder, conducting engineering development flights of its complex weapons systems and man-machine interfaces. Stephenson had helped us during the LM preproposal studies and proposal, and since the contract award had been with LM full time.

To enable Stephenson to manage his expanded role we authorized an assistant for him. He selected Scott MacLeod, a former navy fighter pilot who had flown many Grumman airplanes, including the swept-wing F9F-5 jet. At Grumman MacLeod was a production pilot in the Flight Test Department flying aircraft as they came off the assembly line. Stephenson and MacLeod worked directly with the astronauts on TM-1, forging personal bonds through joint problem solving.

Prior to the TM-1 review we had given considerable thought to the problem of egress to the lunar surface. To help evaluate possible techniques, Rigsby and Sherman designed an apparatus called the "Peter Pan rig," a cable-and-pulley device suspended from the mockup room's overhead traveling crane that counterbalanced five-sixths of a spacesuited subject's weight. This crudely simulated the gravitational force on the lunar surface, which is one-sixth that of Earth. By supporting the subject from a chest harness, belt, and thigh straps attached to the Peter Pan rig and moving the overhead crane to follow the subject's movements, we could evaluate the feasibility and relative difficulty of performing various maneuvers on the Moon's surface.

We evaluated a number of means of getting onto the lunar surface and back into the LM. We thought it could be done by using a knotted rope to climb up and down from the LM and a block and tackle to lower and raise scientific equipment and sample containers. This required a platform on top of the landing-gear support strut in front of the forward hatch upon which the astronaut could stand before descending the rope. This platform was nicknamed the "front porch." Stephenson and Harms tried this maneuver in spacesuits a few days before the review. They found it slow and strenuous but feasible. Objective evaluation was difficult because it took time and practice to adapt to the Peter Pan rig and attain proficiency in moving about with it. We definitely needed NASA's opinion.

During the formal mockup review eighty NASA engineers and astronauts inspected TM-1 and wrote more than one hundred chits, fifty-two of which were presented to the review board. Improvements were made in the feel and positioning of flight controls, equipment stowage within the cabin, the positioning of circuit breakers switches, and other crew compartment internal details. After a full day of evaluation by Ed White, Grumman's proposed use of rope, block, and tackle for lunar surface egress was declared unacceptable. White found it too difficult and unnecessarily hazardous for what should be a routine activity. We agreed to devise alternate schemes and to conduct evaluation sessions on the Peter Pan rig with Ed White and others, aiming at a decision review on TM-1 in May 1964. The cabin lighting was also considered in need of further improvement, especially the electroluminescent panels. A lighting review was scheduled to take place in May (at the same time as the egress evaluation) with the astronauts.

The issue of whether LM needed docking capability at the front hatch in addition to the overhead hatch was examined by the TM-1 review team. Forward docking was a hangover from Grumman's LM proposal but was no longer considered necessary for redundancy. The remaining argument in its favor was LM pilot visibility during the close docking approach maneuver. LM would be the active vehicle during rendezvous maneuvers and approach to the CSM. The LM pilot would fly this segment from his normal flight position, standing in his flight station and looking at the CSM out the front window. If LM could dock with its front hatch upon return from the Moon, the LM pilot would not change his flight position when he handed off active maneuvering command to the CM pilot, who would capture and dock the LM.

I thought it desirable to eliminate the docking requirement from the forward hatch for several reasons. It would save weight by eliminating the forward docking tunnel and docking impact loads on the front face structure, and it would give us more design flexibility with the forward hatch if we needed it to be larger or other than circular. We could probably simplify the hatch-locking mechanism as well. Both Grumman and NASA engineers had suggested that a small rectangular window be placed in the cabin ceiling, directly over the LM pilot's head, to provide visibility of the CM during the close docking approach maneuver.

During the TM-1 review, White, Conrad, and others evaluated this possibility. They tried bending their heads backward to look out the overhead window while wearing spacesuits and restrained at the normal flight station. They noted the difficulty of using the LM controls in this position and of switching eye contact back and forth between the instrument panel and the overhead window. They concluded that the position was somewhat awkward but not unworkable, and it would probably be easier at zero gravity. The astronauts recommended and the board approved a month of study and simulation at Houston before making a final decision, but the overhead docking scheme appeared promising.

TM-1 had most of the electronic equipment mounted externally on the ascent stage behind the pressurized crew compartment. The electronic "black boxes" were mounted on vertical racks facing each other, with a ladder up the side aiding access. During the review a simpler concept of mounting the boxes lengthwise on vertical rails in a single array facing aft was discussed with NASA and accepted for further study.

Several other TM-1 design issues were reviewed with NASA, including the stowage of scientific equipment in the descent-stage bays, the placement of antennas and of handholds to aid crew mobility on both stages, and ground clearance of the descent engine bell. The review stimulated ideas from NASA and Grumman on means of resolving these and other issues. We also used TM-1 during discussions with LM Manufacturing on possible design and manufacturing approaches to building the front face of the crew compartment. It was a flat, pressurized, sheet-metal structure with complex geometry and angles and major penetrations for the forward hatch and windows. I favored welded construction in order to save weight and provide a leak-tight compartment, but the geometry appeared to preclude that solution.

The TM-1 mockup review ended optimistically. Many crew compartment design details had been resolved and others placed on a path to resolution, often using TM-1 as the evaluation test article. Over the next two months, White and Conrad worked on TM-1 and the Peter Pan rig, and with Grumman's Crew Systems, Structural Design, and Mechanical Design engineers came up with a much better concept for egress to the lunar surface. They enlarged and added handrails to the front porch platform and fitted a ladder to the forward landing-gear strut. The astronaut descended to the surface by crawling backward out of the forward hatch, then across the platform and down the ladder while facing it. Returning from the surface he faced the ladder again, climbed up it onto the platform, and crawled forward through the hatch. He could carry a full rock sample container with one hand (at one-sixth G), place it on the front porch ahead of him, and push it in through the hatch. The whole surface egress procedure was made easier, safer, and more intuitive. This design was approved at the interim TM-1 review in May 1964, as were improved cabin lighting provisions and electroluminescent panels, the latter being enthusiastically endorsed by Pete Conrad. The Houston studies of overhead docking also were positive, and the requirement for docking with the forward LM hatch was deleted. Further work by LM Engineering and Manufacturing resulted in the selection of a hybrid manufacturing approach to the front face structure, a combination of welding and riveted construction. With these key decisions we were ready to firm up the LM design configuration and basic operational details with the last planned LM mockup.

M-5

Rathke and I decided to make M-5 an accurate engineering and manufacturing aid that we could use for fit checks, configuration and operations studies, and development of manufacturing techniques. Most of it was metal, with parts made from accurately dimensioned engineering drawings, not just sketches. M-5 was a shakedown for the LM project's drawing release and configuration control systems and provided the newly formed LM Manufacturing Department's first challenge. From our subcontractors and suppliers we obtained accurate mockups of equipment and components, including the rocket engines, reaction control thrusters, environmental control assemblies, tanks, antennas, and flight displays and controls. In some places we were able to install prototype flight hardware, as with the crew station controls, supports, and restraints. We installed flight-type electrical wiring harnesses, connectors, and fluid system plumbing and components. These installations closely simulated the form and fit of flight hardware but were nonfunctional. The external surfaces of the ascent and descent stages were covered with thin, shiny aluminum sheets, simulating the micrometeorite shields that would cover the flight LMs. In its gleaming metallic shell, M-5 looked like an exotic space creature. Did we create this strange apparition or was it built on the Moon by little green men? This LM seemed like it would be at home on the Moon.

More than four hundred Grumman engineering drawings were required to define M-5, plus many more from our subcontractors. In mid-1964 LM Engineering had its first struggle to meet a schedule of drawing releases to Manufacturing; a foretaste of what would soon become our major preoccupation. Bob Carbee personally led the engineering work on M-5, pushing and coordinating the drawing outputs from the LM design groups. The last two weeks before the review Engineering and Manufacturing worked around the clock in two shifts, seven days a week to finish M-5 on time.

There was an embarrassing glitch about a week before the review. When M-5 was moved into position on the mockup room floor, it was evident that we would not be able to demonstrate the deployment of the forward landing-gear strut from stowed to extended position because of lack of ground clearance. During deployment the landing gear extended below the floor if M-5 was supported on the floor by its own landing-gear legs. Raising M-5 up on pedestals under its feet would provide the necessary clearance for the deployment demonstration but would leave M-5 two and a half feet above the mockup room floor, which was not acceptable for conducting lunar egress exercises. We shamefacedly accepted the solution of breaking up the concrete floor and digging a trench through which the front landing gear could sweep during deployment. Engineering took a lot of flak for not foreseeing that problem earlier.

The review commenced on 6 October 1964, with more than one hundred NASA astronauts and engineers in attendance, and continued through 8 October, when the reviewers were augmented by MSC director Bob Gilruth and MSFC director Wernher von Braun and virtually all the Apollo program leadership from Houston. NASA was pleased with the quality of M-5 and its extensive detail, and they closely examined and evaluated each area. Von Braun was very enthusiastic. He climbed up the ladder and through the front hatch into the cabin and inspected the interior. Upon exiting, he called excitedly to a Marshall colleague from M-5's front porch, "You've got to go up there. Go in there; it's great!"[1]

Astronaut Roger Chaffee donned a spacesuit and mockup backpack and practiced entering and exiting M-5 from the surface. He had difficulty squeezing through the front hatch due to repeated hangups with the bulky backpack. He insisted that I watch the maneuver, together with Carbee, Rigsby, and Harms, and afterward declared that the circular shape of the front hatch was wrong. What was needed was a rectangular hatch, somewhat higher than wide, to match the squarish form factor of the backpack. We were convinced by his demonstration and approved the chit he wrote when it came before the review board. Because the front hatch no longer was required to perform docking, the circular shape was not necessary.

The M-5 mockup review board convened with Owen Maynard as chairman and Carbee and I representing Grumman. We sat at a large conference table on the mockup room floor in front of M-5, facing the audience. Microphones on the table and speakers amplified our voices, and we and the chit presenters used a lectern and slides projected onto a large portable screen to address each item under discussion. More than 250 people were present, filling every row of chairs and spilling over into standing room along the walls. Gilruth, von Braun, several astronauts, and other high-ranking NASA officials attended and followed the proceedings with interest, despite the difficulty seeing and hearing the presentations in the crowded hall. A total of 148 changes were proposed, and the board approved 120 of them. Most of the changes were minor and none forced major redesign; even the change in forward hatch shape was readily accommodated.[2] Max Faget was very impressed with our mockup, declaring it was what an engineering mockup should be like. He said North American's mockup was mainly a marketing prop while Grumman's was an engineering design tool.

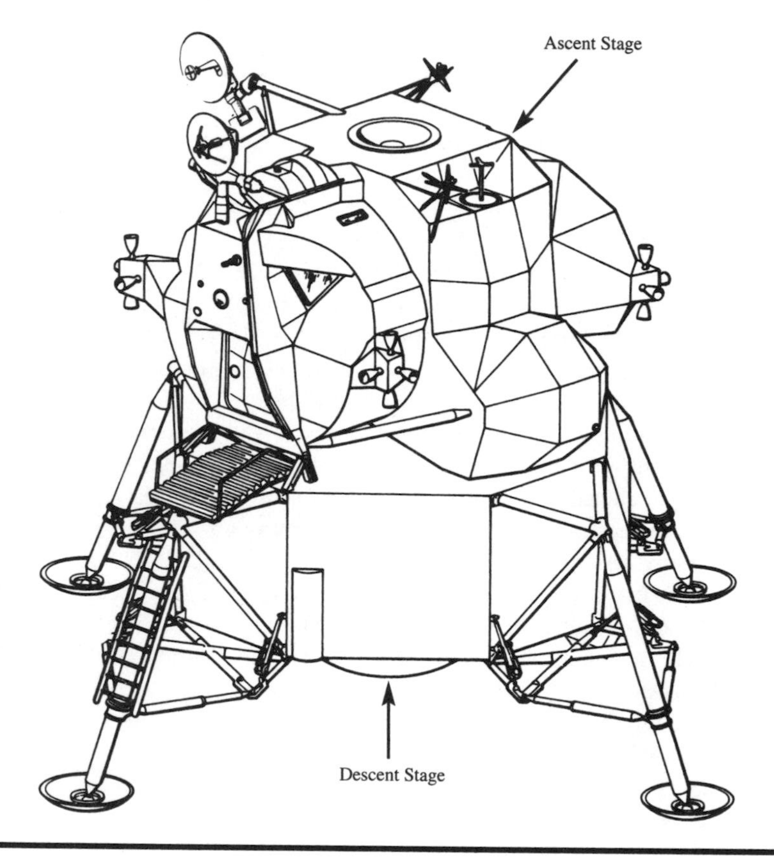

The final lunar module design. (Courtesy Northrop/Grumman Corporation)

The M-5 review provided Grumman management a great opportunity to exchange ideas with our NASA counterparts and to gain insight into their concerns and opinions. Gavin, Mullaney, Rathke, and I all benefited from discussions with Max Faget, Chris Kraft, and astronauts Chaffee, White, and Conrad. Faget was a gifted, intuitive aerospace designer. He made numerous informal suggestions for further improvement or simplifications of designs that were represented on M-5. He was very pleased with the ladder and front porch for lunar egress but, along with Chaffee, recommended we make the hatch bigger and rectangular.

Kraft brought the mission and flight operations perspective to the review. He questioned what the crew would have to do to perform different functions in LM and how the Mission Operations people on the ground could support these events. He challenged us to reexamine various design and operational features for further simplification, making us aware that the crew's time and energy was the most precious commodity on the mission, more than any other expendable, such as propellant, water, power, or oxygen. Our design must conserve this precious resource. The astronauts were very helpful in evaluating and improving anything they would operate or use. Their detailed suggestions for improving the cockpit, instrument panel, flight controls, equipment stowage, docking techniques and aids, and ingress-egress provisions, among other things, contributed greatly to LM's success. Nothing was too minor to escape their scrutiny; their suggestions were practical, simple, and readily implemented.

The M-5 review ended with NASA approval of the LM design configuration. The preliminary design phase of the LM project was over. It was now up to Grumman to bring this full-scale model to life as a functioning, dependable lunar landing spacecraft. My colleagues and I were very proud that we had passed muster by NASA's experts at this comprehensive review. We were also relieved that the required design changes were modest in scope, although we faulted ourselves for not catching the obvious ones sooner.

I packed up my notebook and papers when the M-5 Mockup Board adjourned, and as the crowd dispersed I went up into M-5's cabin for one more look. Then I descended the ladder and walked slowly around the spacecraft, my eyes catching the thousands of details that had to be converted from dummy mockups to reality. The mockup room was almost empty and someone began turning out most of the overhead lights. In the half-darkness I took one more look at the spectral shape of M-5, standing there as LM would one day stand on the Moon, strange, gaunt, otherworldly. We had so much work to do to make it all happen. I hastened to get on with it.

7

Pushing Out the Drawings

With the preliminary design completed the focus in LM Engineering shifted to getting out the drawings, specifications, and other technical-definition documents that would enable Manufacturing to build the LM. The weekly LM program meeting became increasingly devoted to evaluating progress against the drawing-release schedules and recovering from the many slippages that occurred. Program Manager Bob Mullaney bore down hard on Engineering at these meetings and plainly registered his displeasure whenever Engineering failed to meet drawing schedule release promises. This happened repeatedly, to the point where Engineering was losing its credibility with program management. Rathke and I had to correct this chronic problem.

At our weekly project Engineering meeting Bob Carbee and his design group leaders worked over the schedules and reviewed the problems that required solution. In this early stage of detailed design the estimated number of drawings required was very rough, mainly based on aircraft estimating factors considering weight and complexity. The estimated number of drawings required kept increasing, often by hundreds per week, as more details of the design became available. The list of deliverable end items that Grumman would design and build under the contract kept changing as some test articles and rigs were added and others deleted, depending upon changing evaluations of what tests were required to qualify each subsystem. The biggest unknown was the ground-support equipment, for which the requirements could not be defined until the flight hardware that it supported was designed and the basic mission operations and ground test procedures were established. To me the GSE looked like a bottomless pit in which we were becoming hopelessly mired. We needed to do something drastic to clarify this poorly defined area.

The total engineering drawing workload climbed relentlessly. During 1965 and 1966 our estimated total grew from a few thousand to more than fifty thousand drawings, of which more than ten thousand were for GSE. There were also about one thousand specifications and procurement packages required and several hundred test and checkout plans and procedures. The whole program faced a moving target of workload, to the near despair of program management and Manufacturing, who were at the mercy of Engineering to define the deliverable end products before they could prepare reliable manufacturing plans, schedules, and cost estimates.

When the cumulative number of drawings to be released was plotted as a graph against time in weeks or months for 1965 and 1966, it looked like we had a mountain of work to climb. On the plot were layers showing the numbers of drawings required for each category of end item, including the flight LMs, LM test articles and mockups, test rigs, special test equipment, and GSE. The drawing-release schedule graph was displayed on large sliding wallboards in the Plant 25 conference room, where Joe Gavin and Bob Mullaney held the weekly program management meeting. Plotted against the current release schedule were the actual releases, lagging below the scheduled lines, and the latest revisions to the schedule including new promised dates to recover from past slippages and to accommodate further growth in the estimated number of drawings. Revising the drawing-release schedule and tracking our actual performance against it occupied much of the time of Subsystem Project Engineer Bob Carbee and his design section heads.

Late one afternoon a slim, blond, freckle-faced fellow stopped into my office and introduced himself. He was Bill Craft, an assistant project engineer with the E2C program and a former structural designer. Although he had a good job, he was intrigued by the space program and wondered if there was any chance of joining LM and working for me. We talked well into the evening, and our personalities seemed to click. I was impressed with his knowledge and experience with engineering drawing production and working with Manufacturing to resolve problems on the shop floor. I also liked his helpful, frank attitude. "Look," he said, cocking his head to one side and squinting through eyes that were mere slits, "I know what you're thinking—here's another hotshot who's after my job. Well let me assure you that's not the case. There's no way I could even think about doing your job. But I've had experience in getting drawings out and helping the shops build from them, and I think that's where you could use help right now."

After checking with Rathke, Mullaney, and Gavin, all of whom had worked with Craft and thought highly of him, I offered him a job as assistant project engineer. Craft's initial concentration would be on accelerating drawing releases and working with Manufacturing in drawing scheduling and interpretation. He tackled a difficult job with enthusiasm, persistence, and diplomacy and made a major contribution to our ability to schedule and deliver the drawings on time.

Ground-Support Equipment

GSE began holding weekly meetings with GSE Engineering and Manufacturing because of the variety and complexity of their end items and their differences from flight hardware. GSE was a relatively new specialty area in aerospace projects, growing rapidly in importance. Within Grumman, GSE was designed by the Materials and Processes Group of Manufacturing Engineering, which was originally formed to provide tooling and special process equipment and procedures to the Manufacturing Department. It was built either in Manufacturing's tool shop or by the Equipment and Processes (E&P) shop, which handled mechanical and fluid devices. These were small, relatively underfunded groups offset from the mainstream Manufacturing and Manufacturing Engineering activities devoted to flight airplanes.

For the LM project GSE Engineering was assigned to me and Rathke and formed a triad reporting to us, along with Systems Engineering and Subsystems Engineering. Coming from Manufacturing Engineering instead of Aircraft (or Vehicle) Engineering, their heritage and loyalties were very different. The same was true of E&P Manufacturing, which had been integrated into LM Manufacturing. I had almost no prior contact with either of these GSE groups, so I had much to learn and a new cast of characters with whom to establish working relationships.

Several hundred GSE items had been identified, and the list was continually growing. They came in great variety, the most complex being the deliverable GSE end items. These were identified by a five-digit number series and name: for example, "61033 oxidizer servicing cart." After use in LM final assembly and test at Bethpage, these end items were delivered to NASA with the LMs they supported and were installed and operated at Kennedy Space Center or White Sands. (Grumman operated an altitude rocket test facility at NASA's White Sands, New Mexico, complex, where the LM ascent and descent propulsion systems and the reaction control systems were test fired.) Factory support and test equipment was not deliverable but normally stayed within Grumman for use in assembly and test operations. Items in both categories carried four-digit identification numbers: for example, "3022 ascent-stage workstand." There were many miscellaneous items, such as adapters, cable sets, installation kits, and so on, that were often designed or revised at the last minute based upon need. The GSE Publications Group prepared installation and operations manuals for this equipment.

Joe Shea, recognizing that North American Aviation and Grumman were not coping adequately with the challenge of GSE, assigned one of his strongest managers, Rolf W. Lanzkron, to be GSE "czar" for the Apollo spacecraft. Hard driving, dedicated, and knowledgeable, Lanzkron attacked Apollo's GSE problems like a man possessed. With his slicked-back hair, black horn-rimmed glasses, and dark suit, Lanzkron looked more like a lawyer than an aerospace manager. With dynamic leadership, he quickly commanded the respect of the LM GSE people. He was a tough taskmaster, but he was as hard on himself as anyone else. His searching, pointed questions and comments seldom gave personal offense but shone a relentless spotlight on our failings and shortcomings.

Lanzkron transmitted a sense of urgency to all of us. His message was simple: the LM GSE mess must be quickly cleaned up and properly managed or the whole Apollo program would be held up by Grumman's ineptness, with dire consequences for all involved. He held weekly GSE meetings in Bethpage, having just come on the "red-eye" flight from Los Angeles after a similar meeting at NAA the day before. His meeting started promptly at 7:00 A.M.; the normal Grumman starting time of 8:00 was too late for him. With all Grumman's GSE leadership assembled in the straight-backed chairs in the Plant 25 main conference room, Lanzkron sat at the conference table facing them and methodically went through the status, problems, and corrective actions of every end item in the GSE program. A large set of sliding wall boards were prepared and statused, listing the end items and showing their schedule position.

These meetings were long and painful. Lanzkron would not accept waffled answers to his questions. If the person he was cross-examining implicated someone else as part of his excuses, Lanzkron insisted that the third party be brought into the meeting immediately to defend himself or rebut the charges. If information was lacking, he demanded that it be located and presented before the meeting adjourned. The meetings frequently continued until 8:00 or 9:00 P.M. with no break for lunch. When the meeting began Lanzkron prominently displayed on the table in front of him an apple and a glass of water, which everyone knew was his lunch. The Grumman people either brought brown-bag lunches or slipped out in ones and twos for a quick bite in the Plant 5 cafeteria.

After holding his first few LM GSE meetings, Lanzkron requested a management meeting with Gavin, Mullaney, Rathke, and me. He expressed shock and dismay at what he had found at Grumman. Calmly and methodically he presented the data our own people had given him, which showed that we were making no progress against the GSE schedules but steadily falling further behind. His devastating assessment was that Grumman's GSE management was not up to the job and Grumman's corporate GSE capability was inadequate. He demanded strong corrective action from management. Gavin and Mullaney admitted that we were deficient in the GSE part of the program and asked for time to prepare a corrective action plan. Smiling triumphantly, Lanzkron gave them two weeks. He confirmed the time of the follow-up meeting with Gavin before he left the room.

Unknown to me at the time, this pressure on GSE from NASA reinforced a similar thrust from the navy, which required more GSE to support fleet operations and maintenance of Grumman's aircraft and their ever more complex systems. This resulted in a major change in Grumman's corporate organization—the creation of a new operating department for Integrated Logistic Support (ILS). On a level with the traditional departments of Engineering, Flight Test, and Manufacturing, ILS was responsible for every activity and product required to support Grumman's aircraft and spacecraft in their operating environment. This included GSE design and manufacture, spare-parts provisioning, logistic-support analyses, and publications and manuals. The former Materials and Processes (M&P) Engineering and E&P Manufacturing groups were included in the new department, and there was a reallocation of Engineering and Manufacturing floor space to support ILS expansion. To lead this new organization management selected Edward Dalva, a veteran Grummanite and proven project engineer and manager who had directed the development of the W2F and E2A airborne early-warning aircraft systems. Dalva proved effective in infusing energy and a "can-do" spirit into what had been a neglected backwater activity. He greatly expanded Grumman's sales in the ILS area, making it a major segment of the company's business base and increasing our customers' satisfaction with the total performance of Grumman's products.

But LM could not wait for corporate GSE to get its house in order; we needed help now. Gavin and Mullaney moved to strengthen the LM GSE management. Our LM GSE project engineer, Dick Spinner, and the Manufacturing manager, Bob Wagenseil, were simply no match for Lanzkron, often withering pathetically under his onslaughts. John Coursen, a savvy project engineer with a record of accomplishment on the F10F and F111 programs, replaced Spinner, and Tony Oddo, assistant director of E&P Manufacturing, replaced Wagenseil. They recruited other capable people to bolster LM GSE management. Gavin and Mullaney also set in motion a plan to expand and integrate LM GSE manufacturing and test facilities, which were scattered throughout various tooling and E&P shops. In the brief follow-up meeting with Lanzkron, he nodded approvingly when these changes were announced and said he looked forward to seeing what these new people could do.

I spent more time on GSE, often attending at least part of their weekly meeting with Lanzkron as well as continuing to hold my own biweekly technical staff meeting with GSE Engineering. As I became familiar with the GSE program, I soon found myself doing what the GSE "natives" did: thinking of and referring to the GSE end items by their five- or four-digit numbers. In some cases the dialogue at the GSE meetings sounded like the legendary comedians' convention where someone calls out the number of a joke and everyone laughs.

In late 1966 LM GSE Manufacturing was consolidated in a refurbished building in Syosset, Long Island, known as the Davega building for its former occupant, a sporting-goods chain. With the GSE Engineering and Manufacturing staffs greatly expanded and intense concentration and effective leadership by Coursen and his team, plus increased support from the corporate ILS Department, the LM GSE program gradually turned the corner. The end-item list stabilized, and our ability to make schedules and hold to them gradually improved. At its peak in early 1967, LM GSE Engineering and Manufacturing employed more than fifteen hundred people. The total cost of LM GSE was greater than our 1962 negotiated estimate for the entire LM program, an indication of how GSE was underestimated by all at the start of the program.

Lanzkron was reassigned by Shea to key trouble spots elsewhere on the program, but he returned to Grumman for a time in 1967 to help us resolve pervasive manufacturing assembly and quality-control problems. Although his brand of assistance was painful to endure, it was effective and contributed substantially to both Grumman's success and that of the Apollo program.

Configuration Management and Scheduling

The management of the Apollo program was a staggering task, more complex in its scope, number of participants, and interrelations between program elements than anything yet attempted by the aerospace industry. The activities and outputs of more than 175,000 people in thousands of organizations across the United States had to be coordinated, scheduled, and provided with technical interface data wherever their products interacted with someone else's. Interlocking schedules for each program participant had to be prepared and scheduled against actual progress. The whole array of schedules had to be capable of rapid revision whenever new developments, such as unforeseen test failures or delivery slippages, caused a change in technical approach or plans.

As a model of how to manage such complexity, NASA used the ballistic missile programs then under development by the air force and the navy. They adopted major techniques from each of them: from the air force, the configuration management system, and from the navy, the program evaluation and review technique (PERT).

Configuration management, developed and used on the Atlas, Titan, and Minuteman programs, was a formal, structured method for defining and controlling the detailed configuration of aerospace flight vehicles and GSE. Rigorous control over the millions of parts in a missile or spacecraft was essential to prevent unexpected failures due to the actual vehicle differing from what the engineering drawings showed it to be. The leaders of NASA's manned spaceflight program had learned this lesson early, through a terrifying incident that occurred as they all watched the first launch of an unmanned Mercury capsule on a Redstone booster, the last flight prerequisite to a manned suborbital mission. MR-1, as it was called, failed ignominiously and dangerously on the launch pad on 20 November 1960. The Redstone rocket fired briefly and lifted a few inches off the pad, then abruptly shut down. The Redstone settled back on the launch pad, but the Mercury spacecraft continued to go through its preprogrammed sequence of events for liftoff and landing; firing the escape tower rocket, then the radar chaff dispenser, the drogue parachute, and the main parachute. The Redstone was left standing on the pad, unsecured, with its propellant tanks pressurized and no connection to ground control from the blockhouse. Fortunately there was no fire or explosion, and NASA technicians eventually went out to the pad and successfully disarmed the live Redstone rocket.

The cause of this mishap was determined to be a failure of configuration management: the "as built" system was not "as designed" in the engineering drawings. Subsequent to the prior launch from the pad, a technician filed about one-quarter inch from one of the two prongs in the tail plug, an electrical connector at the base of the launch umbilical tower that plugged into the base of the Redstone, and did not tell anyone what he did. When the Redstone lifted off about an inch, the tail plug was supposed to pull out of the base of the rocket, switching the Redstone over to internal power. However, if one of the two prongs was to disconnect before the other, the circuitry was designed to shut the Redstone's engine down immediately. This is what happened when one connector prong had been made shorter than the other. NASA was extremely lucky that MR-1 did not explode on the pad in a spectacular failure that might have caused cancellation of the whole manned spaceflight program. This narrow escape made all those who witnessed it, including Gilruth and von Braun, determined to have a disciplined configuration management program throughout NASA.[1]

The air force's configuration management system was thorough and rigorous and was documented in five manuals known as the "375 series." With minor revisions NASA adopted it completely and educated its contractors on how to apply it. I listened to an air force lecture on configuration management, after which Bill Rector and I discussed how we would apply it to LM. We agreed that the M-5 mockup review would also be the preliminary design review (PDR) as defined by the configuration management manuals, marking the completion of the definition phase of the program. After PDR no major configuration changes could be made without prior approval by the LM Change Control Board (CCB), staffed by both NASA and Grumman. The design and development phase for unmanned LMs would end shortly before LM-1 delivery with a critical design review (CDR), and for manned LMs before LM-3 delivery.[2] After CDR no changes that affected form, fit, or function of any part or component could be made without prior CCB approval. The nomenclature implies the rigor and discipline that this system imposed. Making it happen would require continuous education and repetitive indoctrination of all those on the program, a major task of training and communications. The LM program was Grumman's first exposure to rigorous air force 375 configuration management, as the navy had not yet imposed it on our aircraft programs.

The PERT scheduling and control system was first put into major program usage by Admiral Rayburn on the Polaris Fleet Ballistic Missile program and improved by other users, including the Minuteman program. It was adopted by the Department of Defense as a standard tool of program management, and numerous software programs were developed to provide computer-assisted generation of the PERT schedules and networks. NASA adopted a version of PERT as its standard across all programs, including Apollo. As in configuration management, the LM program was Grumman's first use of PERT, so we had to train our people in the system and its application to LM. During 1964 we conducted many training sessions for our people in the use of CM and PERT.

Many of PERT's features were well suited to the Apollo program. It provided flexible, computer-aided scheduling of program activities and allowed for uncertainties by inputs of "best, worst, and most likely" completion dates for each event. It showed the relationships between events through graphic network diagrams in which the activities or events were positioned and coded to show their relationship to other events, whether as prerequisite constraints or as parallel nonconstraining activities. Using computer software, PERT allowed the program to be scheduled at a low level of detail that could directly be related to measurable actions and outputs in Engineering, Manufacturing, or test. It used the computer to "roll up" the highly detailed network into summary schedules sorted by spacecraft, by test program, or whatever views were meaningful at the overall program level.

Rathke and Mullaney established the LM program Scheduling Group to learn the PERT system and help us apply it. The group was headed by Larry Moran from the Scheduling Group in corporate engineering. Larry was tall, thin, and energetic, with a tanned and freckled face, black hair brushed back in a low pompadour, and a ready laugh and smile. A heavy smoker, he looked unhealthy hunched over the networks and schedules on his drafting board, wreathed in smoke with a cigarette dangling from his lips. He was a natural leader and had a wonderful sense of humor, which helped his hardworking group endure the long hours they put in.

Larry Moran built a group of program schedulers who were far more than just PERT technicians. They understood interactions between program events better than anyone else on LM, and as scheduling problems developed they were very constructive in helping program and engineering management devise "workarounds" to recover lost time or accommodate unforeseen events and requirements. The schedulers roamed freely throughout the LM program, determining the actual status of events scheduled on the PERT networks and posting them on sliding schedule wallboards displayed in the Plant 25 main conference room. Gavin and Mullaney conducted the weekly LM program meeting using Moran and his charts to show the schedule problem areas and discuss recovery options. Summary schedules and actual status were rolled up from our PERT networks and fed into the computers at NASA-Houston for integration into the total Apollo spacecraft schedules and networks that NASA maintained.

Moran's group and their PERT diagrams became the principal tool that Rathke, Carbee, Whitaker, Coursen, and I used to plan and schedule Engineering's activities. The networks included all the drawings and the analyses required from Whitaker's Systems Engineering groups that were constraints on drawings. The flow of major test activities was added as the networks matured. The drawing estimates and scheduled delivery promises came from the leaders of the individual Engineering design sections based upon their past experience, knowledge of the current design, and their workload and staffing. Bob Carbee for the flight and ground test spacecraft and John Coursen for GSE, probed, challenged, and demanded performance against schedule from their people—day after day, week after week. It took time, but the results of their persistent efforts gradually became visible as the growth in our estimated number of drawings required and slippage against the established delivery schedules decreased throughout 1965 and 1966.

Mountains of Work

We faced a mountain of required drawing outputs that built up rapidly throughout 1965 and plateaued in 1966 at more than four hundred drawings a week. The pressure of schedule was relentless. The expansion of our Engineering groups continued in 1965 and 1966, completely filling Plant 25. Engineers worked in large bullpen areas on the open floor with drawing boards and desks jammed five abreast. (To reach the aisle, the person working at the center board had to pass behind and disturb two people on either side.) Group leaders went from board to board several times a day, collecting nearly finished drawings to be turned over to Drafting or the Release Group and updating the status inputs for PERT. Problems involving lack of information from other Engineering groups, or anything that held up completion of a drawing were taken to the group leaders for action or resolution. There was a corner of the first floor of Plant 25 where a large plywood table had been constructed for preparing oversized drawings and "white masters." The table was three feet high and about ten feet square. Several engineers or draftsmen worked on their hands and knees on the table, drawing directly on large sheets of vellum paper or on the white painted aluminum sheets on which the "white master" tooling templates were drawn and later cut out to fit into the tool that defined LM skin contours or structure. The three large Engineering floors in Plant 25 were like throbbing frames in a giant beehive, bustling with purposeful activity, visible output, and the hum of accomplishment.

Plant 35, a two-story building with almost the same floor area as Plant 25, was completed in 1965. It provided some relief to the overcrowding in Plant

25 and space for the continuing growth in Engineering staff. By 1966 LM Engineering also occupied substantial office areas in Plants 5 and 30 as the number of engineers neared its peak of almost three thousand. Art Gross and the Facilities Management Group were busy with office moves, expansions, and reconfigurations, which had to be carried out with a minimum of disruption to the ongoing work.

The Drafting Group was a veteran operation in corporate Engineering, dating back to Vice President of Engineering Dick Hutton, a draftsman-engineer who was the first person hired by Chief Engineer Bill Schwendler when Grumman was founded. Draftsmen took drawings after the engineer had drawn and specified the basic design and added the finishing touches, such as dimensions, notes, and additional views, that were needed to make them intelligible and useful to manufacturing. They created detailed parts drawings from assembly drawings and engineering instructions. They did preliminary work to create a drawing by preparing cross sections or views upon which the engineer could work out his design additions or modifications. We often had more than four hundred draftsmen on LM, a number that could be readily be changed by bringing in "job shoppers," contract draftsmen from manpower supply companies. Extensive use of overtime in Engineering and Drafting evened out the peaks and valleys in manpower demand.

Corporate Drafting was run by Howard Krier and the LM Drafting Group by Ross Chandler. They were a hard-bitten pair of aircraft design veterans and not about to take any nonsense from a young project engineer like me, or from an upstart government agency like NASA for that matter. Krier was a fatherly looking fellow with salt-and-pepper hair and clear-rimmed glasses who considered himself the final authority on anything pertaining to drafting. He had a high-pitched voice that became increasingly annoying to me in debates when he would stubbornly refuse to give ground. Chandler, black-haired and usually toting a large cigar in his mouth, was somewhat younger and had the patient but exasperated look of one who believes he has seen it all.

I had one meeting with them in which, in response to pressure from Shea in Houston, I challenged the schedule status and manpower estimates of all LM Engineering, including Drafting. They unrolled dozens of detailed estimating sheets broken down to individual drawings and went through them page by page while Krier kept up a high-pitched skirl of justification and Chandler blew cigar smoke over my head. By the time they were finished I was glad to escape with an estimate no higher than their original position. After that I relied on Rathke to deal with them—he understood them better and, if necessary, could blow cigar smoke back at Chandler.

The LM Release Records Group was led by Al Caramanica, a bright young graduate from Drafting. He had the task of transforming the group to carry out far more complex functions than it had in the past. Release records on aircraft programs was a mundane "go-fer" and record-keeping activity, which obtained and recorded the necessary approvals on drawings from the analytical engineering groups, such as Stress, Thermo, and Loads, and released them to Manufacturing and to the customer. They also issued and recorded all the engineering orders, which were changes affecting released drawings. With the advent of rigorous configuration management on LM, the Release Group became the implementer and enforcers of the configuration control system. They handled each drawing and EO as required by NASA's regulations, making sure that CCB approval was obtained before releasing a controlled drawing. The Release Group became a major resource for the CCB, providing the manpower needed to enforce configuration control discipline in LM Engineering and working closely with the Quality Control (QC) Department, which provided surveillance and enforcement elsewhere on the LM program.

The ultimate purpose of the engineering drawings was to provide design information to LM Manufacturing from which they could make and assemble the spacecraft and its GSE. Manufacturing was under heavy schedule pressure from program management and needed to have the drawings released to them in sequences that would permit efficient buildup of production activities in the shops. Close cooperation between Engineering and Manufacturing was imperative. Frank Messina, manager of LM Manufacturing and a seasoned veteran of aircraft production since World War II, and Bill Rathke had worked closely with each other on several airplane programs; they both knew what was needed. A joint Engineering-Manufacturing production planning committee was established to coordinate Engineering's drawing releases with Manufacturing's needs. Initially headed by Bill Craft of Engineering and Bill Bruning of Manufacturing, this group met daily, comparing Engineering's plans and ability to release drawings with Manufacturing's requirements and preferences on the sequences of releases. The result was a "negotiated" release schedule that gave Manufacturing the best available sequences and kept them informed in detail as to what was coming "down the pike."

In especially tight schedule situations, in which days and even hours counted, we modified our basic drawing-release system, taking calculated risks to save time. We permitted the use of advance prints, which had not yet been analyzed and approved by Stress, Loads, or other analytical engineering groups, on time critical drawings. We told Manufacturing what the risks were and what aspects of the drawing might change after the analytical review, letting them decide whether the time saved was worth the risk of change. I was never comfortable with that subterfuge and tried to limit its use.

Unfortunately, the natural order in which design drawings are produced is opposite from Manufacturing's needs. A designer visualized the whole unit or assembly he was designing, producing sketches and an assembly drawing showing how the device fits together. From this the individual detailed parts making up the assembly were pulled out and drawn up, complete with the dimensions, material specifications, and instructions that allowed the shop to make them. Manufacturing naturally wanted the detailed parts drawings first, so they could obtain the raw materials and purchase components and then fabricate the parts. With all the parts in hand they needed the assembly drawings to put it all together, although Manufacturing Engineering needed the assembly drawings early to plan and design the assembly tooling. The production planning committee did their best to accommodate these conflicting priorities.

As 1965 stretched into 1966, there was no relief in the unending schedule pressure on the LM program. Because LM was the last major Apollo program element to be defined and placed under contract, Grumman started a year behind everyone else. NASA's George Mueller and Joe Shea concentrated special attention on Grumman's ability to meet schedules, and they were not at all satisfied with our performance.

Shortly before the M-5 mockup review, Gavin, Mullaney, and I were summoned to Houston for a review of LM management's performance by Joe Shea and Bill Rector. Shea told us that NASA planned to have the first Earth orbital flight test of the CSM on a Saturn 1C booster in February 1966, and they wanted LM's first flight to be in February 1967. Given Grumman's performance so far there was no hope of achieving that goal. Grumman was demonstrating a 0.8 "schedule slip ratio": every five weeks our schedule promises slipped by four weeks.

"Your management performance in other areas has been no better," Shea said in an exasperated voice. "Your cost estimates have already doubled since negotiations and are still rising. The LM's weight grows weekly with no sign of leveling off. By every measure, your performance is extremely bad. We had hoped Grumman would outperform North American—it looks like you are even worse. You are the people we're counting on to make Grumman perform. If you can't do it, we'll have to ask Towl and Titterton to find someone who can." I shuddered inwardly at this bare threat. Shea was at the end of his patience with us.

Shea recounted a list of serious deficiencies in Grumman's management of the LM program. He said we had encountered unnecessary problems through management ineffectiveness. Examples were late procurements, slow and incomplete GSE identification and delivery, and sluggish staffing buildup. He faulted our subcontractor relationships, traceable to fractionalized divisions of responsibility within Grumman, which prevented us from executing a "total package procurement" approach and establishing teamwork with our major subcontractors. Most grievous of all, Shea accused Grumman of failure to understand or accept NASA's Apollo program philosophy. He cited many examples of this, such as our recurring disagreements over reliability: whereas NASA's approach was careful design and thorough ground testing to disclose weaknesses and verify fixes, Grumman, based on its major GNC arguments with MIT and other debates, seemed wedded to a statistical approach. We seemed to dispute NASA's selected approaches to black-box maintenance (ground only), qualification test requirements (test to failure not always necessary), and manned-flight prerequisites (based primarily on ground test program). We made little effort to find common-usage equipment from North American Aviation in areas like GSE, reaction control, and communication, despite repeated NASA prodding.

Finally Shea faulted us for an arrogant attitude: "Grumman is a proud organization and you are proud people. Yet your box score on technical decisions has not been good." He found Grumman reluctant to admit to a problem without having the solution in hand, which we expected NASA to rubber stamp. He cited, for example, our switch to riveted construction on the LM's front face without telling NASA—for months they thought it was all welded. Similarly, Grumman was far along with a two-tank ascent propulsion design before NASA knew we were considering it—the same when we went to three fuel cells from two. In every case NASA found that Grumman had failed to perform analyses that were necessary to determine whether these changes should be made. He urged us to talk out technical problems with NASA as they arose so they could jointly contribute to the solution.

It was a humiliating dressing down, but Shea ended the meeting on a positive note, outlining a series of steps to correct these shortcomings and reestablish a team relationship between NASA and Grumman. These consisted of a series of meetings in each of the recurrent problem areas where the differences in philosophy and approach would be thoroughly aired and resolved.[3]

I was shaken by that meeting. I felt that much of it was directed at me personally and most of the problems lay in my areas of responsibility. Some of the perceived differences with NASA did not even reflect my own views; for example, I strongly favored the pragmatic design-and-test approach to reliability over statistics, and I was a major proponent of NASA's decision to eliminate in-flight maintenance. That meant that I had not been effective in explaining and enforcing my own concepts within LM Engineering at Grumman. The indictment of our poor schedule performance was true, as I saw week after week at the project meetings, where despite the explanations and excuses, the bottom line result remained terrible. As for arrogance, I tried hard to avoid it myself and never failed to reprimand my people when I heard them referring to NASA or others in a demeaning manner. I felt we were newcomers to manned spaceflight and had no basis for a superiority complex.

After the M-5 review, Rathke, Carbee, Whitaker, and I bore in on the drawing production problems. In the weekly project meetings and my daily technical staff meetings we looked for constraints to drawing output and took actions to eliminate them. A new wave of Engineering staff expansion was approved and implemented. Although we had been satisfied at our former staffing levels, Shea's review of our engineering shortcomings showed that we had much more work to do. The glamour of the Apollo program helped us attract talented engineers from throughout the country. We expanded into Plants 35, 5, and 30, and we applied overtime and added job shoppers. The production planning committee intensified its efforts to support Manufacturing; this primarily meant developing credibility to Engineering's promise dates.

There was progress from all these efforts but it was slow and not steady. Unforeseen problems and complications confounded our attempts at orderly recovery, and increased definition and understanding of the program resulted in longer lists of end items to be delivered. It was like fractal geometry—the closer one viewed the task ahead, the more detailed tasks were visible. At times the amount of work ahead seemed unassailable; the pile was always growing. There was a mounting sense of frustration within Engineering: the schedule pressure was unrelenting, and no matter how much we did, it was never enough. Gradually our drawing output climbed, from fifty a week to one hundred, then two hundred and still climbing. We needed more than four hundred drawings a week to meet the program's requirements.

NASA's unhappiness with our performance attracted Grumman corporate management's attention. We began seeing more of Senior Vice President George Titterton, Chief Technical Engineer Grant Hedrick, and Director of Flight Test Corwin H. ("Corky") Meyer.

Hedrick was a gifted engineer who had joined Grumman as a stress analyst during World War II when the bridge designing he was doing for consulting engineers Parsons and Brinkerhof was suspended for the duration. He impressed Bill Schwendler and Dick Hutton with his keen ability to analyze aircraft structural designs and his careful application of basic engineering principles. His wartime structural design work on the Tigercat, Bearcat, and Albatross earned him the post of chief of structural engineering in 1947. In that position he succeeded Bill Schwendler as the unofficial "chief blacksmith of the Grumman Ironworks," the guarantor of the structural integrity of Grumman's airplanes. He later became chief technical engineer, in charge of aerodynamics, structural analysis, and structural design. Widely respected throughout the aerospace industry, Grant Hedrick was "Mr. Engineering" within Grumman.[4]

Hedrick was of medium height and solidly built, with sandy hair and wise-looking eyes behind rimless gold eyeglasses. A highly ranked amateur tennis player, he was trim and athletic. He grew up in Fayetteville, Arkansas, and graduated from the University of Arkansas with a degree in civil engineering. His country background showed in his practical, no-nonsense approach to engineering in which he concentrated on the facts and avoided unfounded theories or speculation. He was demanding of his subordinates, insisting that they study their problems carefully and perform thorough analyses of alternatives before coming to conclusions. Those who did not meet his strict standards seldom got a second chance. The structural engineering community at Grumman looked up to him as their leader and the arbitrator of safe airplane design. I did not know him very well, and at first I was somewhat intimidated by his reputation and his crisp manner of cross-examination.

Hedrick's increased involvement with LM Engineering was quite helpful. He set up a regular weekly meeting with the Engineering managers and section heads that I attended with any LM Engineering support I required. Both sides could select agenda items for discussion. These were useful meetings aimed at insuring that LM was making use of the best talent available at Grumman to solve technical problems. Hedrick had a remarkable ability to cut through complex arguments or voluminous calculations and find the hidden flaw that rendered the whole argument suspect. After NASA's senior structural analyst, Joe Kotanchik, watched Hedrick in action at Grumman, he enlisted him to serve on an ad hoc task force assembled by Marshall, Houston, and their contractors to solve a serious problem, known by the undignified term "Pogo," on the Saturn booster.

Pogo vibrations occurred along the long axis of the booster, when fluctuations in thrust by the rocket engines fed back into increased pressure at the pump inlet, reinforcing the magnitude of the initial disturbance. It was feared that uncontrolled Pogo vibrations could destroy the Saturn. This phenomenon only occurred in flight, where the increased thrust produced increased acceleration and hence pressure at the base of the propellant tanks. With its cause understood, it could be modeled in computer analyses. Grant Hedrick was effective in guiding NASA's analyses toward a practical solution of the Pogo problem. In a notable meeting at Houston with Kotanchik and the structures experts from NASA, Hedrick spent almost an hour studying an array of strip chart recordings and reduced data from the latest Saturn flight test that were displayed on tables in a large conference room. Asking questions as he went, he pondered its significance. Then he returned to one set of data and pointed out numbers that appeared unreasonable and required further checking.

Hedrick was right, and his astute observations put NASA on the path to a proper analysis. Later he would apply his unique talents to the solution of some of LM's most difficult technical problems.

The involvement of Meyer and Titterton was sometimes less beneficial. Meyer was generous in making available talented Flight Test managers to the LM program. These men had valuable experience in real-time scheduling and dealing with unanticipated problems while conducting challenging test programs. But Meyer's generosity sometimes came with a string attached. He maintained "back-door" communication with some of his former people on LM and sometimes used this to his own ends with corporate management, as when he mounted an internal campaign to remove the checkout and test responsibility from the LM program and assign it to the Flight Test Department.

George Titterton's contributions also had pluses and minuses. On the plus side, he had the internal clout to get the LM program what it needed, in budgets, people, facilities, and equipment. When convinced, he moved swiftly

NASA officials view the lunar module mockup at Bethpage. *Left to right:* Joe Shea, Tom Kelly, Bob Gilruth, and Joe Gavin. (Courtesy Northrop/Grumman Corporation)

and effectively to place added resources at our disposal. This was extremely valuable during the buildup years of 1965 and 1966. He was also a hard-driving, ramrod manager who set high standards and accepted no excuses, and he quickly developed a close rapport with Joe Shea because in this regard they were soul mates.

On the minus side, Titterton's personal management style could be divisive and confrontational. His established method was to pit the Engineering,

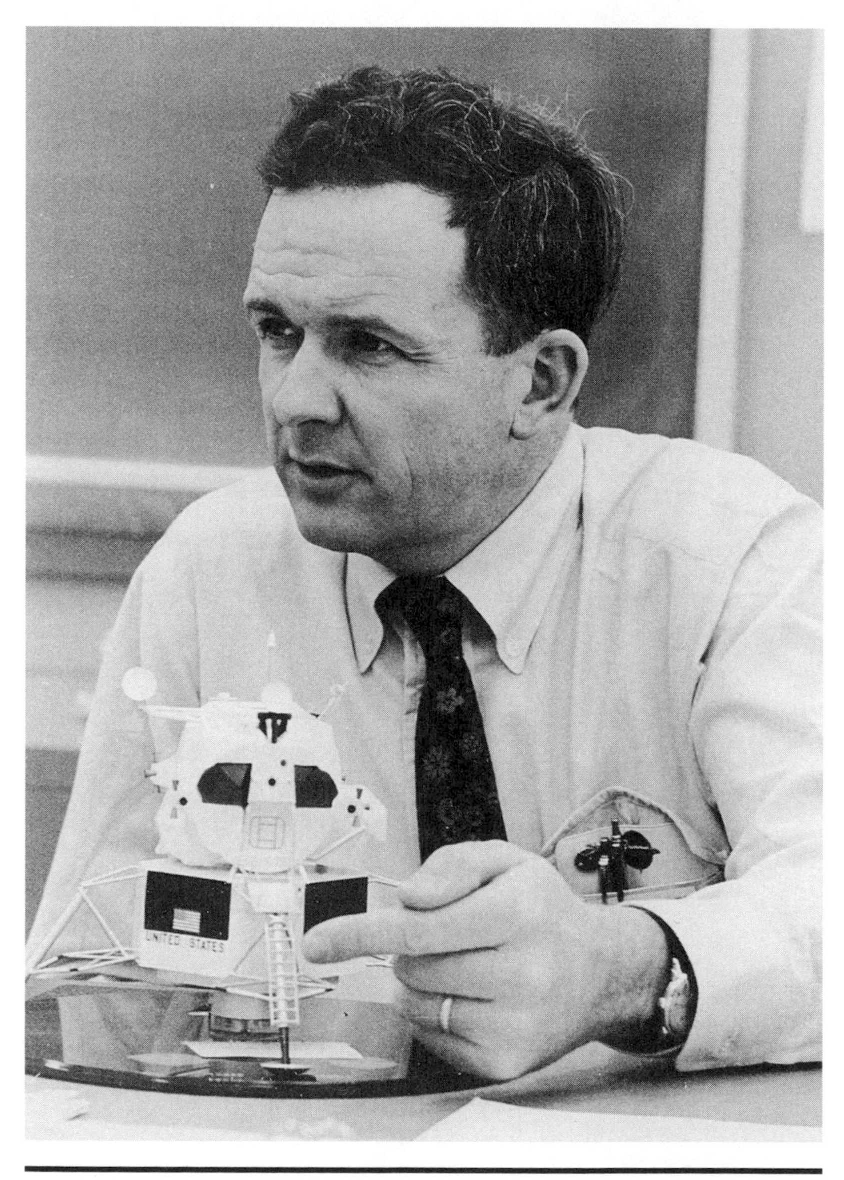

Tom Kelly in his Bethpage office, 1965. (Courtesy Northrop/Grumman Corporation)

Manufacturing, and Flight Test Departments against one another, showering all sides with heavy doses of sarcasm and ridicule. Outside of work he was engaging, witty, and considerate, a perfect gentleman, but when exercising his command authority on the job before an audience he could be a difficult man.

Titterton enjoyed being the champion of the long-suffering shop workers who were continually misled and frustrated by the out-of-touch ivory tower engineers ("knuckleheads," he called them). I tried to counteract his attitude because I believed that open, honest cooperation and teamwork between Engineering and Manufacturing was our only hope of reaching our goals.

By early 1966 LM drawing production had hit its stride; on 18 April I noted that 556 drawings had been released in the prior week, 356 of which were for the flight vehicles and 200 for GSE.[5] We were finally making our schedules, but it had been a long struggle, more than a year and a half, and we were now faced with a new challenge: getting the LMs through final assembly and test. This would require the development and validation of several hundred test plans and operational checkout procedures (OCPs), documenting every detail of the tests so they could be programmed into the ACE stations and witnessed by Grumman and NASA Quality Control. No sooner did we reach the required high plateau in drawing output than another mountain to climb loomed in the near distance.

Just Another Job?

Under such a continuous grind, program excitement wore thin. LM was becoming more like a job, with such high pressure that it exacted a high toll from its participants. We had almost succeeded in taking the fun out of building a spacecraft to take men to the Moon. But not quite. Every month the full Moon reminded me of the scope of our grand endeavor and made all the pressure, deadlines, schedule commitments, and management in-fighting worthwhile. I could withstand Shea's wrath and Titterton's ridicule because I knew that when our LM landed the first astronauts on the Moon's pristine surface, these travails would count for nothing. Perhaps like childbirth, the pain would be forgotten in the glory of a new beginning.

I was insulated from any thought that we might not be successful. By reducing the LM program to thousands of PERT "inchstones"—minor events that could be scheduled and verified when they happened—we had removed any doubt that the conclusion would not be successful. By following the "yellow brick road" of our PERT networks, we were sure to reach the Moon. The only question was when, and would I still be on the program with my reputation intact when we got there?

8

Trimming Pounds and Ounces

Even before Grumman was placed under contract the weight of the LM had grown, and as we began the actual design of the craft, its weight continued to increase at an alarming rate. Our LM proposal design was estimated at 22,000 pounds, but in discussions with NASA at the contract negotiations it was agreed to increase LM's weight to 25,000 pounds. The target weight, which was used for propellant-tank sizing, was increased to 29,500 pounds in February 1964. This was the basis for the tank sizes in the TM-1 and M-5 mockups. The estimated weight of the design soon exceeded the target, and we again faced the need to increase the target weight and the size of LM's propellant tanks.

Because of the flight mechanics and rocket propulsion constants of the lunar mission, LM was extremely sensitive to weight growth. For every pound in the ascent stage, three pounds of propellant were required to land it on the Moon's surface from lunar orbit, lift it back into lunar orbit, and rendezvous with the CSM. The weight growth factor was, therefore, 4 to 1: every pound of added ascent-stage inert weight increased the Earth launch weight of the LM by four pounds. Weight added to the descent stage but left on the Moon had a growth factor of 2.25 to 1. This was far greater than for aircraft, even for military missions with fighters and attack bombers, where growth factors seldom exceeded 15 percent, or 1.15 to 1.

The major factors that drove LM weight up during 1963 and 1964 were reliability requirements, mission operational requirements, and configuration definition. The reliability approach that we had adopted relied upon functional or component redundancy and ample safety factors where redundancy was not possible. This was a sound design philosophy, but it added weight compared to our proposal design, which had little redundancy.

Operational requirements became visible to us through the work of the Apollo Mission Planning Task Force. Their design reference mission provided the basis for estimating many operational requirements that affected LM weight: the duty cycle (on/off times) of LM equipment requiring electrical power, the number of cabin pressurizations with oxygen, the spacecraft attitude relative to the Sun and hence heat loads and cooling water capacity, the amount of consumables (water, oxygen, power, propellants) required on LM, and so on. There was a wealth of data in the DRM that had an effect on weight. As we dug into this, our understanding of mission operations requirements improved and the weight grew.

Configuration definition changed swiftly in 1963, when the ascent stage's geometry was completely revised in the M-1 mockup. A year later the M-5 mockup essentially completed the definition of LM's basic configuration. The weight estimates were revised based on this geometry and equipment arrangement, and invariably they were greater than before.

An empirical explanation for the large early weight growth on LM lies in the historical data accumulated on many aircraft and spacecraft programs, which shows that estimated weight, from design sketches and system descriptions, is typically 20 to 25 percent lower than weight in the final product, mainly due to the omission of many components and design details that are unknown to the estimators. Calculated weight, from engineering drawings, is usually 5 to 10 percent low because many small details, such as the exact number of fasteners, cut-outs, and installed components, are only approximated in the calculations. Not until actual weight is obtained by putting the various parts and assemblies on a scale is the bias toward low estimates corrected. As the LM design moved from sketches to drawings to fabricated parts, the weight increased with the increasing percentage of calculated and actual weights versus estimated.

No matter how we rationalized it, the inexorable growth in LM weight was threatening the whole Apollo mission. Caldwell Johnson was concerned that LM might become too heavy to do its job. Johnson's opinion was not to be taken lightly. He had designed the Space Task Group's own versions of the LM and constantly reviewed and critiqued Grumman's emerging design, looking for improvements, simplifications, and weight savings. I obtained sound guidance from him, Maynard, and Faget in evaluating LM design choices.

In mid-October 1964 Carbee, Whitaker, and I had a lengthy meeting with Owen Maynard, Caldwell Johnson, and William A. Lee of the Apollo Operations Planning Division in NASA-Houston to review the LM's weight status and tank resizing options. The NASA people had done their own analyses of LM's weight growth history and status and found the outlook bleak. They projected 12 to 15 percent growth in LM's weight from its then-estimated 30,200 pounds, which would exceed the maximum allocation for LM based upon a Saturn 5 boost capability of 32,000 pounds. They suggested possible LM design changes to reduce weight and modifications to the mission rules and trajectories that might allow for an increase in the fraction of the Saturn 5 payload weight allocated to LM.[1] We had a month to study these suggestions and add to them, and to recommend a target weight for LM resizing and an improved weight-control approach.

We responded that LM should be resized to a target weight of 32,000 pounds at Earth launch, of which 10,800 pounds would be in the ascent stage. This fully utilized the allocated Saturn 5 payload for LM. We also recommended a number of weight-reduction items, including the use of supercritical helium instead of gaseous helium to pressurize the descent propellant tanks, reduction of LM hover time at landing from two minutes to one, and possible use of batteries instead of fuel cells for electrical power. Batteries might not save weight, but they promised great simplification of the electrical power system and increased reliability. We urged further study of this option. Also recommended for study was replacement of the LM rendezvous radar with star trackers or lasers. Mission changes that we suggested NASA consider included use of a non–free return trajectory to the Moon and reduction of the CSM's midcourse correction propellant allowance, both of which would allow more of the Saturn 5 payload to go to LM. After extensive discussions, these recommendations were approved by Maynard, Lee, and Johnson, then, the next day, by Joe Shea. Shea made clear to me that this was all the weight allowance they had to give LM, and that I had better get control of the weight growth or the whole program would be in deep trouble.[2]

Rathke and I preached reduction and control at all our Engineering meetings and set up a system to identify and evaluate potential weight-reduction items. We devoted more time at the daily technical staff meetings to reviewing and deciding whether to incorporate these changes into the design, subject to approval by NASA and the Change Control Board. Still, the monthly LM weight-status report showed increases, and Grumman and NASA managements grew restive. Some of the largest increases were in equipment supplied by our subcontractors and vendors. In March 1965 Bill Rector met with me and strongly urged Grumman to set up an intensive weight-reduction effort, with emphasis on subcontractors and with continuous involvement with and direction by Grumman's LM program office.[3]

Outside events overcame my single-minded dedication to LM: on 27 March 1965 Joan gave birth to a beautiful eight-pound baby boy. He was our fifth son and sixth child, and we named him Peter. I took two days off to mind our other children and take Joan and Peter home from the hospital. Not without pangs of conscience, because I was in the middle of finalizing the design of the newly adopted battery EPS and also the accelerating weight-reduction and drawing-schedule efforts. I soon disappeared into the vortex of work but I took my turn with middle-of-the-night baby feedings.

Shortly after Rector's request for more vigorous and effective weight reduction by Grumman, Arnold Whitaker presented a plan for tightening control on subcontractor and vendor equipment weights. He recommended that we include the allowable weight-change limit as part of every change request that Grumman issued to its suppliers and require the subcontractor to estimate the weight impact as part of his change proposal. This would be followed up by specific accounting of the actual weight impact of the change. The subcontractor's weight-control performance would be a factor considered in determining his incentive fee. Gavin and Mullaney promptly implemented Whitaker's suggestion.[4] However, as the reports came in from this tightened control over the subcontractors, it was clear that LM again was increasing in weight. By July 1965 we were reporting LM as exceeding the control weight of thirty-two thousand pounds.

Rathke and I decided that an all-out "crash" effort was needed to reduce LM weight and to control further growth, despite the disruption this might cause to our equally critical drive to meet drawing-release schedules. We initiated a "scrape" program, led by Sal Salina, head of the Weight Control Section, Len Paulsrud, head of Vehicle Design, and Dick Hilderman, head of Structural Analysis, to whittle weight from the LM structure. By making more extensive use of chemical milling, precision machining, and material substitutions, we could obtain significant weight reductions without major changes in the structural arrangement of LM. Unfortunately, this was not enough for the twenty-five hundred to three thousand pounds of reduction we needed as a margin against further growth.

From Grant Hedrick we learned of a successful approach Grumman recently used on the F111 aircraft program that we could apply to LM. Grumman was a major subcontractor to General Dynamics, supplying the aft fuselage and tail assemblies for all F111s and the design and final assembly of the navy version of this joint air force–navy airplane. General Dynamics–Grumman had won the F111 contract within weeks after the LM award. Despite a very extensive pre-award preliminary design and mockup, as the detailed design proceeded, both the air force and navy versions outgrew their weight limits and engine capabilities. To save the F111 program, which was threatened with cancellation if its weight and performance goals were not met, General Dynamics–Grumman mounted a drastic response, called the Super Weight Improvement Program, or SWIP. Even though most of the airplane and system designs had been completed and released to manufacturing, an independent team of SWIP design experts was turned loose on the program to second guess every aspect of the design for weight savings. The SWIP team worked cooperatively with project Engineering, but they had a direct reporting channel to the F111 program manager to assure that none of their recommendations would be stifled. A dollars-per-pound criterion was established to set a threshold for accepting weight reductions; for the F111 it was five hundred dollars per pound. SWIP was successful on the F111: weight growth stopped and there was some net reduction in weight. The cost of the changes was high but affordable, and the resulting impacts on the schedule could be endured. SWIP saved the F111 program.

Early in 1966 Roy Grumman retired as chairman of the board and was replaced by Clint Towl. Llewllyn J. "Lew" Evans was then appointed president of Grumman Aerospace Corporation. Lew was a lawyer who came to Grumman as chief counsel after working in the Legal Department. He had a flair for marketing and deal making and became vice president of Business Development, strengthening and expanding Grumman's close working relations with the navy. A charismatic and inspiring leader, Lew lost no time in getting acquainted with Grumman's newest major customer, NASA. He held internal reviews on the status of the LM program and was alarmed by the array of problems he found involving our technical, schedule, and cost performance. To provide direct corporate-level oversight of the program, he established the Executive and Technical Review Board (ETRB), chaired by Senior Vice President George Titterton. At its first meeting the ETRB urged Gavin to conduct a SWIP on LM and offered to make available the SWIP review team, which was then finishing its work on the F111.

In July 1965 I was put in charge of the LM SWIP activity. The Grumman SWIP review team, twelve engineers headed by Ed Tobin and his deputy, Paul Wiedenhaefer, would report to me for the duration of the exercise. They had unlimited access throughout the program and a direct reporting channel to Grant Hedrick. NASA management liked the plan; they were pleased that Grumman was taking forceful action to control weight. Bill Lee was assigned by Joe Shea to be my counterpart as co-chair of the SWIP team.

Anyone could make weight-reduction suggestions to the SWIP team; opinions were actively sought through the employee suggestion program. Grumman was responsible for evaluating SWIP items and making recommendations to NASA for approval. We held intensive weekly reviews with NASA, usually at Bethpage, which Bill Lee always attended, often accompanied by Maynard, Johnson, and other engineers. Internally the SWIP team met several times a week with LM Engineering and Manufacturing management, reviewing hundreds of SWIP weight-reduction items and suggestions and thousands of Grumman and subcontractor drawings. Each item was evaluated and dispositioned against the criterion, which after considerable study and discussion had been set at ten thousand dollars per pound for "round-trip" items. Not every SWIP item could make it in time for LM-1; some of the more difficult items were phased in at LM-3, LM-4, or even LM-5. Even so the disruption to the schedule was severe and required constant replanning and revision of the PERT networks and schedules as we fought to maintain forward motion on the program while accommodating some major design changes to save weight.

Tobin and Wiedenhaefer were thorough, persistent, and innovative in finding areas to reduce weight. They were curious about how every system worked on LM and how the critical loads, safety factors, and materials choices were arrived at, so they led us through a complete review and justification of the LM design criteria and choices. Some LM engineers bristled at being second-guessed on their designs, but Ed and Paul were so logical in their questions and approach that no one could take offense. At their suggestion, we began holding SWIP meetings at our major subcontractors' facilities, too, as more than half our SWIP items were in subcontractors' equipment.

Although the SWIP team worked full time implementing the weight-reduction effort, I supplied the technical leadership and made the decisions for Grumman. Joe Shea, Joe Gavin, and I kicked off the SWIP effort at a motivational meeting with a large audience in the Plant 25 main conference room, making clear that the future of the Apollo program could very well depend upon the success of our efforts. We introduced Tobin and Wiedenhaefer to the LM people, and they told briefly what they had done on the F111 program and how successful it had been. They were confident of being able to do it again for LM. I made clear that all LM Engineering managers would be heavily involved and would be accountable for delivering their portion of the required weight savings. In the active question and answer session that followed, our engineers clearly showed their desire to support this program but expressed reasonable concerns and asked for direction in how to balance conflicting priorities in their workloads. We promised to expand the SWIP guidelines beyond the dollars-per-pound criterion to include schedule and reliability criteria also.

My daily technical staff meetings and weekly SWIP meeting were the management focus of the effort. The SWIP meeting usually took three or four hours. Tobin and Wiedenhaefer went through the list of potential SWIP items, noting new additions and reviewing the status of each. Whenever an item was ready for decision, whether to implement, modify, or delete, the SWIP team would be joined by the "cognizant" LM subsystem engineer (called the "cog engineer") or section head.[5] Bob Carbee attended all SWIP meetings, contributing to the decisions and following up in implementing them in his subsystem design groups. At the technical staff meetings the SWIP team usually was not present, but we examined in detail the design issues involved in one or two SWIP items in the subsystem under discussion.

We started the SWIP effort by establishing target weights for each subsystem, broken down into all its parts and components. These were arrived at in reviews with the cognizant subsystem engineer, his Engineering section head, and representatives from his principal subcontractor and suppliers. Drawings and specifications for the subsystem and its major components were reexamined and avenues of possible weight reduction identified. Because most of the designs had been released to manufacturing and many of the system components had been through some amount of development testing, we also discussed the potential impact on schedule and cost of these changes. The outcome of these meetings was that the cog engineer, his section head, and his subcontractor accepted a weight-reduction "target to beat" for SWIP.

I concentrated on the "big-ticket" items that had major potential for weight reduction, although no item was too small to escape the fine net trolled by the SWIP team (items down to .1 pound were considered). Structure headed the big-ticket items because there was so much of it. Redesign, scrape, and materials substitutions were possible in virtually all parts of the LM structure, and thanks to the inspired and diligent efforts of the Vehicle Design Section we saved every possible ounce. The Materials Section played a key role in identifying substitute lightweight materials and in perfecting the chem-milling process in Manufacturing, which was a major technique for reducing the weight of structural parts.

Implementing supercritical helium pressurization of the descent propellant tanks was another big weight saver. Feasibility of this item depended upon the design ingenuity of the LM Propulsion Section and their cryogenic tank subcontractor, Airesearch Division of Garrett Corporation, and by the Fluids GSE Section and their cryogenic tank and component supplier Beech Aircraft. Neither I nor NASA were willing to approve the supercritical helium change until we saw a design for the GSE that convinced us that it would be practical to load and unload this hard-to-handle material on the Apollo launch pad at Kennedy Space Center.

I also gave major attention to the substitution of batteries for fuel cells in the electrical power system. This change was made to improve reliability because the use of batteries eliminated a very complex system of hydrogen and oxygen tanks, plumbing, and components—as well as the fuel cells themselves. From a SWIP standpoint, I had to assure that minimum weight increase resulted. We had to be very sure of what we were doing, both in verifying the battery suppliers claims and accurately assessing the weight of the fuel cell system. Our fuel cell subcontractor, Pratt and Whitney, was well along in building and testing development units both for LM and the CSM, which used a similar system. Grumman held a competition between battery suppliers Eagle Picher and Yardney, and obtained electrical test results and actual weights from both of them. On 26 February 1965 Shea approved the change to Eagle Picher batteries for LM.[6]

The environmental control subsystem was a considerable source of weight savings. With the cooperation of its subcontractor, Hamilton Standard, the ECS Section was able to wring dozens of pounds out of this vital and complex subsystem. After studying many possible configurations, changes to the oxygen-supply section were adopted that optimized it for minimum weight. Cryogenic storage of liquid oxygen was investigated, but we selected a simpler system with high-pressure (2,730 psia) storage of gaseous oxygen in the descent stage. Two much smaller low-pressure tanks in the ascent stage provided the oxygen needed from lunar liftoff to rendezvous. Additional weight savings were gained by structural redesign in titanium of the truss assembly that housed and supported many of the ECS components located inside the LM cabin, and by shaving weight from the components themselves.

As the LM weight concerns intensified, pressure grew to reduce the layers of functional redundancy that provided LM and CSM with lunar-orbit rendezvous capability. In 1964 the system had rendezvous radars on both LM and the service module, but in February 1965 the SM radar was deleted and an optical tracking light added to LM. Shortly thereafter, as a result of the study we recommended in our November 1964 weight-reduction response, Cline W. Frasier of NASA suggested replacing the rendezvous radar in LM with an optical system as well. Consisting of a star tracker in the LM, a xenon strobe light on the SM, and a hand-held sextant for the LM pilot, the optical system claimed reductions of ninety pounds and $30 million.[7] We were directed to make the LM compatible with an optical rendezvous system, and in August 1965 AC Electronics was chosen to develop the optical tracking light.

Late in the year, Mueller, Shea, and Robert C. Duncan set up what they called a "rendezvous sensor olympics" to be completed in the spring of 1966. The intent was to conduct laboratory demonstrations of the RCA rendezvous radar and the optical system and compare the performance capabilities of each. Before the olympics even started the astronaut office indicated a strong preference for the radar because it was more flexible and self-contained. The radar directly provided the critical rendezvous parameters of range (distance to the target) and range rate (velocity of closure with the target), while the optical system derived range from the VHF (very high frequency) communications system and did not provide instantaneous range rate.[8]

In June 1966, after tests and presentations by competing contractors RCA and Hughes Aircraft Company, Grumman recommended that the RCA rendezvous radar be retained and the LM-active optical system dropped. NASA's Sensor Olympics Review Board agreed with the radar choice. The optical system no longer offered cost savings over the radar; as its development proceeded the optical system's cost estimates had grown to approximately equal the radar's. The optical system was still lighter, but by mid-1966 the SWIP effort had been so successful that this one item was no longer crucial. NASA's decision was strongly influenced by the adamant position in favor of the radar taken by astronauts Slayton and Schweickart. Gemini rendezvous flights had been successful using rendezvous radar yielding on-board range and range rate, and the astronauts argued that Apollo should build directly upon these recently demonstrated experiences and procedures.

SWIP was painful and expensive. Besides the great disruption it caused to our schedules and manufacturing buildup, it forced us in Engineering to approve designs and processes that we had concerns about. I felt torn between two opposing interests: the need to reduce weight versus the importance of safety, reliability, and maintainability. I never compromised on safety, and rarely compromised on reliability, but I frequently took chances on reduced maintainability, which would come back to haunt me during the two long years of ground assembly, test, and checkout that the flight LMs endured.

Probably the worst choice I made from a maintainability standpoint was the use of fine 26-gauge wire and miniature connectors on signal circuits, which carried voltage but essentially no current. LM had many miles of such wiring, so this one item saved hundreds of pounds—but at the cost of recurring wire breakage and difficulty of mating and demating electrical connectors in the spacecraft. Upgrading to a higher strength copper alloy for the later vehicles (LM-4 and subsequent) helped alleviate the breakage problem.

Grumman's Manufacturing Department had become expert in chemically milling aluminum alloy and other metals, and we applied this process liberally to structural sheet metal and machined fittings. With chem milling we could create pockets and shapes that were impractical to obtain by conventional machining, and engineers were able to optimize their designs so there was almost no excess metal remaining except that required along the load paths. We obtained major weight savings from this technique. Unfortunately we later found out that chem milling rendered some alloys and part shapes more susceptible to stress corrosion. This occurred when the parts were assembled with less than precise fit-ups into the mating structure, which locked in stresses when the fasteners were tightened, and were exposed to a humid, salty environment, which was standard both on Long Island and at Kennedy Space Center. Months after the first chem-milled parts were installed we encountered cracks, and many inspections and reworks resulted at Bethpage and Cape Kennedy.

With supercritical helium the penalty was system complexity, both onboard and in the GSE, plus a total lack of operational experience by NASA, the military, and the equipment suppliers with such a system. Supercritical helium had to be kept just a few degrees above absolute zero (–459 Fahrenheit), requiring the best available insulation and vacuum technology and extreme care in the fabrication of all components of the system. On Apollo 9's launch we found out just how tricky this could be.

There were many other changes, some of which caused concern, if not actual problems. We made more extensive use of titanium instead of steel in applications where this would save weight. Titanium was harder to shape and machine and more expensive. It was also in short supply, and even with the Apollo program's high priority we encountered some delays with large titanium forgings.

We chem milled the skin of the crew compartment down to exactly the thickness needed to meet the structural safety factor criteria and no more. This yielded a thickness of .012 inch (twelve-thousandths of an inch) for the cylindrical aluminum alloy wall of the crew cabin, or about as thick as three sheets of household aluminum foil. The wall could easily be damaged by a dropped tool or heavy foot, and was therefore covered with protective hard plastic liners throughout its long period of assembly, test, and checkout at Bethpage and Kennedy.

In addition to the big-ticket weight-reduction items, weight was saved by detailed scrutiny of every system and design in the LM. Some of the weight came out in ounces, but it added up to worthwhile amounts. Substantial weight was saved by redesign of the micrometeorite shielding and thermal blankets that covered the external surfaces of LM. We made a very efficient and lightweight combination consisting of fiberglass plastic standoffs bonded to the underlying structure, supporting a .005″ aluminum alloy sheet of micrometeorite shield, and insulation blankets consisting of multiple layers of aluminized mylar only one-eighth of a mil (.000125 inch) thick. The mylar blanket material was specially developed for this application.

In the electronics area, emphasis was given to lightweight packaging to save weight. Wire wrap terminals within the electronic boxes were used to increase reliability and reduce weight by minimizing the number of internal friction connectors. Short, direct conductive heat transfer paths, from the heat dissipating components on the printed circuit cards to the attachment flanges of the electronic boxes, were used to enhance thermal control and reduce packaging weight.

In the fluid systems, weight and complexity were minimized by using pressurized propellant or fluid feed wherever possible, with relatively low pressure levels, small pressure drops, and small forces. The requirements for system safety and reliability were paramount, but the weight considerations affected the details of the design. The propulsion and reaction control systems had propellant and pressurant tanks of lightweight titanium, and stainless steel interconnecting plumbing lines that were joined by high temperature brazing.

All this painstaking effort paid off handsomely. The growth of LM weight was abruptly arrested in mid-1965, and a reduction of more than twenty-five hundred pounds was achieved. Reductions continued throughout 1966 until actual weights began to dominate and the changes resulting from the LM critical design review in March 1966 were felt, showing further increases. A similar weight-reduction effort on a smaller scale was conducted late in 1967 and provided an acceptable margin below the spacecraft control weight. For the later missions (Apollos 15–17, LMs 10–12) increased Saturn 5 capability permitted an increase in LM weight at Earth launch to thirty-six thousand pounds. The four-thousand-pound growth was used to provide an extended lunar stay time of three days and to carry the lunar roving vehicle, a collapsible, electric-powered car for traversing the Moon's surface, and additional scientific equipment and sample return capability.

Participation in the LM SWIP effort was very widespread in Engineering

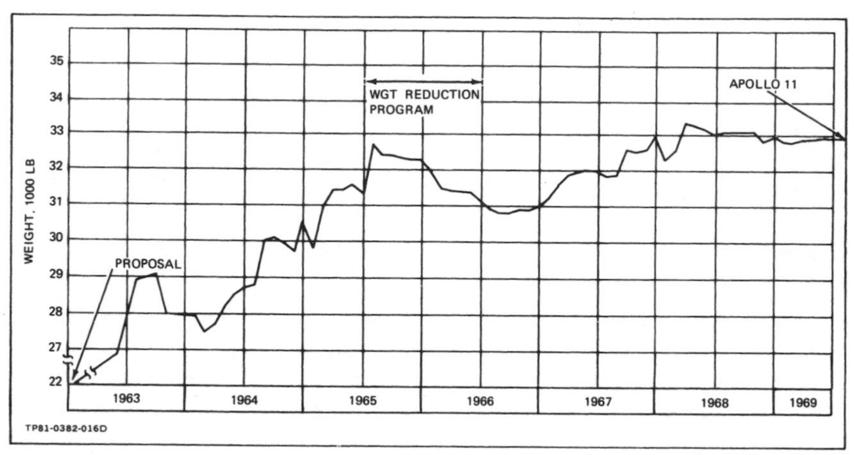

Fig. 16 LM Weight History

The lunar module's weight history from proposal to Apollo 11. (Courtesy Thomas J. Kelly)

and Manufacturing and throughout our major subcontractors. The difficult work of devising and searching out possible weight-reduction candidates was spearheaded by Tobin and Wiedenhaefer and the SWIP team. Their persistence and drive inspired us all and was reinforced by LM Engineering management, especially Carbee, Whitaker, and their section heads. The whole exercise also benefited greatly from the steady support and guidance provided by NASA's Bill Lee, Owen Maynard, and Caldwell Johnson. They were firm in demanding results but optimistic about the outcome.

I was very upset at having to tear apart the LM design as it was nearing completion because I also felt the intense schedule pressure from NASA and Grumman management. Phase-in of changes provided a practical relief valve to the schedule impact problem. In the beginning I also wondered if we could achieve our weight-reduction goals, but the supreme confidence shown by our SWIP team and their NASA counterparts overcame my doubts. Within Grumman it was a great team effort; all elements of the program contributed where they could. Manufacturing suggestions often resulted in process improvements or design simplifications that saved weight. In engineering, the design and materials groups were particularly important. Some new materials, like the ultra-thin aluminized mylar insulation, were created in response to industry stimulus by our Engineering Materials Group.

Near the end of 1966 SWIP activities took up less of my time as the effort successfully wound down. By then a new round of problems and priorities had arisen to claim my attention and LM weight was a less urgent concern. I felt quiet satisfaction with the results of SWIP, even though we could not claim any particular credit for it as we were just fixing the weight growth problems that we ourselves had generated. Grumman rose to the challenge, however, and provided the techniques, leadership, and skilled people who solved the problem within the cost and time constraints of the program. We could not have done otherwise.

9

Problems, Problems!

As the LM moved from paper design to hardware from mid-1966 to the flight missions in 1969, the work of Engineering gradually shifted from design to trouble shooting and problem solving. The philosophy of the Apollo Project was to obtain reliability by tracking down the root cause of every test failure and correcting it. This idea was reinforced at Grumman, where we persistently examined the data, analyzed, and retested until we found the reason for each test failure. As more LM components and assemblies were tested, more failures occurred, and we found ourselves tracking hundreds of failures that could not be closed until we had found the cause and implemented corrective action.

The failures were recorded by Quality Control, which witnessed all tests, and dispositioned by the Reliability Section in Engineering. Many different groups within Grumman were often involved in analysis, retesting, and devising appropriate corrective actions. George Weisinger, head of LM Reliability, led the failure tracking and closeout activities. A thin, wiry fellow who seemed to be constantly in motion, Weisinger knew where to look for likely causes of failures and whom to enlist in finding a solution. At the peak of the failure reporting and corrective action effort in 1968, we averaged three hundred open LM failure items "on the books" and had several hundred engineers at Grumman and our subcontractors engaged in closing them out. Each failure was followed through the closeout process to assure that none were "lost in the shuffle."

Most of the failures were tracked down and resolved promptly—within a few weeks and without major cost or schedule impact. A few were chronic, recurring problems that required technical detective work to find the underlying cause and design ingenuity to devise an effective solution. Every LM subsystem had major development problems, so the severity of the problems was a matter of degree and depended upon the point in the LM program at which the problem arose. Problems that emerged later in the program, closer to flight launches, were more difficult to handle without major impact to the program schedule.

In this chapter I recount some of the more memorable technical problems, chosen because of either their impact on the program or the extent of my personal involvement in their solution. This is only a small sample; for LM to reach the Moon and return, *all* problems had to be closed out prior to launch from KSC.

Propulsion and Reaction Control System Leaks

Leaks in these critical systems were a chronic problem from the completion of the first heavyweight propulsion test rigs (HA-1 and HD-1) in mid-1964 through the manufacture of the flight LMs. Leaks occurred on both the pressurant (helium gas) and the propellant fuel and oxidizer sides of the system. The rocket fuel was called 50/50 (50 percent hydrazine, 50 percent unsymmetrical dimethyl hydrazine, or UDMH), and nitrogen tetroxide (N_2O_4) was the oxidizer. Although no leaks were tolerable anywhere in the systems, propellant leaks were extremely serious because the propellants were highly toxic volatile liquids, and being hypergolic, their fumes would ignite if combined. The search for leak-tight joint designs was long and arduous on LM, and improvements were still being made well into the flight program.

I was concerned about leakage in LM's fluid systems even before we prepared our proposal. Without working out the details at that time, we proposed using welded or brazed joints to eliminate mechanical connections in fluid systems and mechanical energy absorbers in the landing gear to eliminate the potential leakage of hydraulic shock absorbers. As the LM design took shape in 1963 and 1964 we chose stainless steel tubing joined by high-temperature nickel-silver brazing for the propulsion and RCS systems to minimize the number of mechanical joints. Threaded or bolted mechanical joints with gaskets or O-ring seals were allowed for replaceable components, such as tanks, valves, filters, and regulators.

This design policy looked good on paper, but as soon as we gained experience with real system components and plumbing in the cold flow facility at Bethpage, where we pressurized and flowed the systems with inert fluids, and the "live" propellants and hot firing facility at White Sands, it was revealed as inadequate. None of our mechanical joint designs were leak-tight and even the brazed joints leaked unless they were perfect, with full, even flow of the brazing material over the contact area of the joint.

We detected leaks by pressurizing the system with air or nitrogen containing a small fraction of helium and using a sensitive helium detector ("sniffer"), a portable mass spectrometer set to detect a single element, helium. The sensitivity of the sniffer could be adjusted from one to fifty parts per million; the pass/fail setting was two parts per million. When helium exceeding the detection threshold was encountered the sniffer sounded an audible alarm horn, which blew louder the more helium present. The first time we checked the heavyweight rigs for leaks at low pressure on the factory floor it sounded like an air raid was in progress, with sirens wailing everywhere. For a while we blamed the sniffers (too sensitive!), but we found that many of the suspect joints would also show leaks in the common, low-tech bubble test with detergent solution. We stopped the leaks on the heavyweight rigs at Bethpage by replacing gaskets and O-rings and tightening bolts and threaded fasteners to their allowable torque limits.

When the rigs were pressurized at White Sands, leaks recurred, and test crews continually struggled to tighten and seal leaky joints. As they progressed to live propellant loading tests, leaks became hazardous—and visible. The rocket fuel produced a visible white vapor from the smallest leak, and the oxidizer vapor was reddish brown. Both were toxic to humans at the level of a few parts per million. White Sands technicians had to don SCAPE (self-contained air protection equipment) suits to perform the cleanup, decontamination, and repair. The oxidizer was able to find the smallest leak, barely above the threshold of the sniffers,[1] and to expand the leak by corroding the metal when flowing at high velocity. White Sands contended with this situation as best they could while reporting the problem repetitively in their communications with Bethpage.

By January 1967 Lynn Radcliffe, Grumman's base manager at White Sands, could stand it no longer. He made a special trip to Bethpage to demand that propulsion and RCS be made leak-tight, no matter what it took. Two additional sets of rigs, heavyweight HA-3 and HD-3 and lightweight PA-1 and PD-1, had been delivered from Bethpage to White Sands, and although they had improved mechanical joint designs, they still leaked. The lightweight rigs had stainless steel tubing, and some of their brazed joints leaked too. Radcliffe was alarmed and furious. He thought the LM program and Grumman's reputation were headed for disaster unless the leaks were stopped quickly.

Radcliffe was tall, handsome, and articulate, a big man with the lithe grace of a trained athlete. A highly competitive long-distance runner at Syracuse University, he continued to train while at Grumman, frequently running ten miles from work to his home in Lloyd Harbor. This was long before jogging was a popular activity: one night a suspicious Bethpage policeman brought him to the station house for questioning, sure that anyone running along the road must be running away from a crime scene. He also competed in cross-country ski races in the White Mountains of New Hampshire, holding his own against not only other Americans but also athletes from Finland and other Scandinavian countries. He loved sailing on Long Island Sound, challenging the elements in a thirty-five-foot 1906 Herreshof-designed sloop with grossly excessive sail area that was always teetering on the edge of capsizing in strong winds. An athlete, outdoorsman, and engineer, he threw himself wholeheartedly into anything he undertook.

Radcliffe was a flight test engineer before joining the LM program. He worked with the design engineers and the test pilots to develop the flight test programs to validate and certify Grumman's experimental airplanes. He flew aboard the test flights, recording gauge readings from the instrument panel as the test pilot maneuvered the new airplane through its paces. He had developed a healthy skepticism regarding engineers' claims for their designs: "Well, we'll see how good it is when we test it." It was the perfect attitude for one charged with proving that LM's rocket propulsion systems were safe and reliable.

His flight test work had shown Radcliffe how good Grumman's products were—the best carrier-based military aircraft in the world. He was fiercely proud of the quality and excellence of Grumman airplanes; to him the "Grumman ironworks" and "as sterling on silver" were articles of faith not to be profaned.[2] A minister's son, he saw the LM program heading for an apocalyptic end unless it reformed and returned to the high standards set by the company's founders.

Radcliffe started with Joe Gavin and Bob Mullaney and worked his way down through the LM management hierarchy, preaching against the evil of leaks and the need to reform our designs without delay. When he reached me I was shocked by what I heard. I had not realized that the rigs were still leaking badly despite the improved seal designs we had provided. Carbee and his fluid systems design group leaders joined the meeting at my request, and after Radcliffe repeated his message of warning we discussed what to do next. Radcliffe's opinion was unequivocal: "Eliminate *all* mechanical joints, and learn how to make brazed joints that don't leak." We agreed to work toward this goal.

In the ensuing weeks we eliminated the AN (army-navy standard), Gamah, and other threaded fittings that were used with replaceable components. Instead we brazed these components directly into the system with high-temperature nickel-silver brazing alloy, using a tubing stub provided on the component. For replacement, the Manufacturing Engineering group worked out a technique of cutting out the component and using the induction brazing tool to heat the stubs on the component and the plumbing side of the joint to remove the two halves of the cut braze sleeve ("de-plucking" was the descriptive if inelegant term the shop gave to this operation). The stub ends were cleaned and polished and the new component was brazed into the system using the standard techniques. This made maintenance of these fluid systems more difficult and time-consuming, but if we made the brazes properly, they did not leak. It was necessary to X-ray every brazed joint to assure that full penetration of the brazing alloy had been achieved with minimum void area. If it had not, the joint had to be reheated with the electrical induction heating tool using slightly longer time and higher temperature than the previous attempt. Up to four reheats were allowed in trying to get a good joint—beyond that, the joint had to be cut out and resized, a new braze sleeve had to be installed, and the brazing process was repeated. Portable X-ray equipment was used to inspect the joints in place on the shop floor, but no one could work in the immediate area while X-raying was in progress.

There were still some areas where we retained mechanical connections, mostly on the propellant and pressurant tanks. Large openings in the tanks were necessary for cleaning, inspection, and installation of quantity gauging sensors. I retained bolted flange designs in these areas, and at the tank manufacturers' plants we found such designs to be leak-tight.

Even Radcliffe found the improved system design to be acceptable, except that he occasionally had to reheat brazed joints or tighten bolted flanges that had developed leaks. Once he was conducting a high-level NASA delegation, headed by Wernher von Braun, through a tour of Grumman's test facility at White Sands. It was the Huntsville team's first visit, and they were very interested and asked many questions. When they visited the descent-stage altitude chamber to observe a flight-weight test rig loaded with live propellants, the test conductor briefed them in front of a portable projection screen, standing with his back to the test rig. As he spoke, Radcliffe, who was in the back of the chamber facing the test article, suddenly saw an ominous white cloud rising from the rig. He immediately cut the speaker off and shouted, "Gentlemen, there's a fuel leak. Get out now—follow me!"

The Huntsville group moved with alacrity, following Radcliffe into the adjacent control room. The altitude chamber door was actuated shut, safely containing the fumes. It was embarrassing, but thanks to Radcliffe's quick action nobody was injured or even exposed to the deadly propellant fumes.

Leaks continued as an occasional nuisance item in cold flow and at White Sands until the LM-1 fiasco in June 1967. LM-1 was delivered in the midst of shakedown problems with the spacecraft assembly and test operation in Plant 5, Bethpage. When we delivered LM-1, Grumman and the local NASA inspectors thought we had a leak-tight spacecraft, based upon passing helium sniffer tests while pressurized with helium at a low pressure of 50 pounds per square inch (psi). (Operating propellant tank pressures were between 175 and 235 psi, different for each of the three systems.) Soon after it was received at KSC, LM-1 was found to have widespread leakage in the propulsion and RCS systems. The people at Cape Kennedy quickly characterized Grumman's first flight-worthy spacecraft that we had proudly, if tardily, shipped as a "piece of junk that leaked like a sieve." It was a hard blow to the morale of the people at S/CAT, who had put in long hours and seven day weeks to deliver LM-1 and were convinced that it was a quality product.

We initially thought that the cape's findings were due to differences in leak detection procedures and equipment. To test this hypothesis a QC crew from

Cape Kennedy came to Bethpage and performed leak tests on systems in LM-2 and LM-3 that had been found leak-tight at Bethpage. Discouragingly, they found some leaks that had escaped detection by the home team. Moreover, the leaks were real—on both LM-1 at the cape and LM-2 and LM-3 some of the leaks detected by the sniffers could also be seen in the bubble test. Although I was unsure why this happened, I declared that we would adopt the cape leak test regimen to the letter and have experienced cape inspectors train our people in its use.

It took three weeks to standardize our equipment and complete the Cape Kennedy procedures training at Bethpage. The cape used a different model helium sniffer than we, although made by the same manufacturer, Veeco. Embarrassed and responding to pressure from NASA, Joe Gavin became directly involved in the leak problem. At my recommendation he put Will Bischoff, deputy Structural Design Section head, in charge of an intensive leak fix effort. The bolted flanges on the tanks were the worst problem, followed by the few other smaller mechanical component connections that had survived my earlier purge.

Bischoff consulted with the tank manufacturers, Aerojet (ascent propulsion), Allison (descent propulsion) and Bell Aerosystems (RCS), and with O-ring and sealing experts around the country, and developed a new design for the tank flanges. It had dual O-rings, revised groove dimensions and tolerances, and a test port between the two O-rings to detect leakage. Test samples performed very well, as did the first tanks with the new flange design. After an all-out effort by the tank manufacturers, we replaced the tank flanges in LM-1 with an interim improved design which was compatible with the existing tank side of the joint.

Bischoff's team also developed an improved dual O-ring flange design for pipe-mounted components. We used this on components that were themselves problems and required frequent replacement, such as the pressure regulators. The better-behaved components were directly brazed into the system.

Working with Manufacturing Engineering and Grumman fluid systems engineers from Cape Kennedy, the Bischoff team went over our brazing processes thoroughly and upgraded them regarding cleanliness, dimensional tolerances, and prebrazing sizing of the tubing and braze sleeves to assure dimensional accuracy. Brazing and X-ray crews were given additional training, delivered in part by the manufacturers of the induction brazing and portable X-ray equipment. With time, the percentage of first-time acceptable and leak-tight brazes increased and the number of reheats went down.

LM-1 was finally made leak-tight at the cape after three months of intensive effort, aided by six of Radcliffe's best "leak fixers" on loan from White Sands. The damage to Grumman's reputation was severe. The next spacecraft we delivered, LM-3, was pounced upon savagely when it arrived in June 1968 and immediately checked for leaks. This time we had invited the Cape

Kennedy receiving inspection team up to Bethpage to join our people in the predelivery inspections and tests, so the cape people agreed that LM-3 was leak-tight when shipped. Two minor leaks were found in the cape receiving inspection and were quickly repaired. (However, LM-3 had more than one hundred other deficiencies, some of them major, including stress corrosion and wire and splice problems. Nineteen areas were selected for consideration by George Mueller's Certification Review Board before LM-3 could be cleared for flight. The unavailability of LM-3 for flight in 1968 prompted George Low to devise an alternative lunar-orbit mission with CSM alone—Apollo 8, flown in December 1968.)[3]

Propulsion and RCS leakage remained a concern throughout the duration of the LM program. Constant vigilance and retraining were required to attain leak-tight systems—any minor slip would soon be shown up by a squealing sniffer in S/CAT or at Cape Kennedy. The frequency of leaks was greatly reduced from the mortifying debacle of LM-1 or the constant problems that had bedeviled Radcliffe at White Sands, but the occasional leakage that did occur reminded us constantly of the difficult and unforgiving nature of pressurized fluid systems in space. If it leaked in space or on the Moon there would be no way to stop it or replenish the precious lost propellant.

Ascent Engine Instability

Compared with the familiar internal combustion engine with its rotating crankshaft, camshafts, pumps, injectors and reciprocating pistons, valves and pushrods, a rocket engine seems relatively simple. The LM engines were simpler yet, especially the ascent engine, which sought extremely high reliability by straightforward design and rugged construction. Designed to perform a single continuous burn for about seven minutes at constant thirty-five hundred pounds of thrust, the ascent engine had neither pumps, igniters, gimbals, nor a fuel-cooled nozzle bell. It was simpler than the common oil burners used for home heating. Standing four feet high and thirty-one inches in diameter at the nozzle exit, it did not even look formidable, despite the typical rocket engine profile of cylindrical combustion chamber and bell shaped nozzle. Yet this innocuous-looking device proved to be one of the greatest threats to the Apollo program schedule. Frequently the LM ascent engine made the notorious "show stoppers" list as a problem that could stop the enormous, nationwide Apollo program dead in its tracks. When that happened, Mueller, Gen. Sam Phillips, and other NASA leaders applied pressure to Grumman that made life miserable for Lew Evans, George Titterton, and Joe Gavin, and they shared their unhappiness down the chain of command. As hard as we all pressed, ascent engine combustion instability was a chronic problem that yielded only slowly and grudgingly to trial-and-error solutions.

Combustion instability is the most fundamental technical problem in a liquid-propellant rocket engine. Rocket engine combustion requires that the fuel and oxidizer propellants flow into the combustion chamber at a constant rate; be thoroughly mixed in a fixed ratio of fuel to oxidizer, depending upon the particular propellant combination, and be maintained in a confined volume upstream of the nozzle throat (smallest cross-sectional area) for a sufficient time for the chemical reactions of combustion to occur. This is typically done by forcing the propellants under pressure through an injector, a metal plate with many precisely dimensioned holes (orifices) that meter the flow to achieve the desired rate, and direct the streams of fuel and oxidizer to impinge upon one another at a fixed point away from the surface of the injector. The injector resembles a high-flow showerhead—water is typically used in the first checks of the injector's flow rate and stream impingement patterns.

Combustion instability can occur because the rate of energy release in the combustion in the chamber is extremely high, and many physical and chemical variables interact there, each capable of influencing the others. Combustion chemistry, chamber pressure and temperature, injector flow patterns, propellant flow rates, chemical and acoustic energy can all perturb one another in the chamber's roaring inferno. Geometrical tolerances in the injector, bubbles in the propellant, variations in propellant supply pressure, and acoustic and thermal shocks at startup can further act to initiate or sustain instability. Acoustic pressure pulses bouncing back from the chamber walls can bend some of the liquid streaming from the injector orifices, causing more or less rapid combustion and generating new waves of pressure. Instability results in large amplitude oscillations in chamber pressure and heat transfer into the chamber and nozzle walls, which can increase uncontrollably until the engine explodes or ruptures. Typically instability does not occur every time a rocket engine is fired—it's a statistical phenomenon, with instability failures occurring on average once every X number of starts.

Combustion instability has been a problem in rocketry since von Braun and his people encountered it at Peenemünde during World War II. It was usually fixed by making ad hoc changes to the injector and the combustion-chamber geometry, changes arrived at by "cut and try" on each new engine design, an approach that did not provide any reliable rules for the next design. When it was encountered on the mighty F-1 engine for the Saturn S-1C first stage, which stood fifteen feet tall and produced 1.5 million pounds of thrust, it received unprecedented attention. An F-1 engine destroyed itself due to combustion instability in a test stand at Edwards Air Force Base on 28 June 1962, setting off a feverish round of design changes and tests led by Jerry Thomson of NASA Marshall and Paul Castenholtz of Rocketdyne, the F-1's builder.

At first the effort was frustratingly slow, partly because the random nature of the instability did not produce the problem often enough or in any predictable manner. When it did occur, the instability destroyed a huge F-1 engine—an expensive way to learn you still have the problem. After losing two more engines in early 1963, Thomson and Castenholtz devised a technique of exploding a small bomb (like a blasting cap) inside the chamber of a firing engine and observing how quickly the pressure oscillations triggered by the bomb damped out. Arbitrarily they decided the engine would be considered stable if the oscillations damped out within 400 milliseconds, that is, .4 seconds. The bomb test technique allowed the engineers, not nature and statistics, to initiate instability on the test stand. They were ready to shut the engine down quickly if the oscillations diverged rather than damping out, thus avoiding the loss of an engine in their tests.

The heroic efforts of Thomson and Castenholtz in slaying the dragon of instability on the F-1 is another story well worth reading.[4] They did not succeed until early 1965, when the F-1 engine was qualified after more than two and a half years of struggling with this intractable technical problem. In their wake they left the Apollo program management extremely sensitive to the gravity of rocket engine instability, and they also left a well-developed bomb test technique.

Although we and our ascent engine subcontractor, Bell Aerosystems, were aware of instability problems on the F-1 and other contemporary engines, we thought our engine's small size and simple injector pattern might render it immune to that affliction. The initial test firings at Bell in 1963 and early 1964 went well, until NASA realized that Grumman had not imposed bomb stability test requirements on Bell. The ascent engine was derived from the successful Agena engine, designed for the air force's unmanned spy satellite program, for which there were no bomb stability requirements. NASA said this was not good enough for a manned spacecraft engine and we somewhat shamefacedly agreed—it had been an oversight on Grumman's part.

The first bomb stability tests in mid-1964 showed a problem—the combustion chamber pressure oscillations triggered by the bomb did not damp out. They did not diverge either, just continued at constant amplitude for the duration of the firing. The bomb did not always cause undamped oscillations, and even when it did the engine seemed to be no worse for the wear after the test. Manning Dandridge, our LM Propulsion Section head, and Dave Feld, ascent engine program manager at Bell, were puzzled by this uncommon engine behavior, and they consulted widely with combustion instability experts in NASA, the air force, and industry. No one had seen exactly this instability signature before (steady, undamped oscillations), but all agreed it was unacceptable and must be eliminated.

For the next two years Bell tried every instability cure that they, we, or NASA could think of, to no avail. Dandridge's usual cheerful optimism wore thin; he grew increasingly anxious and frustrated. At first supremely confident that the next baffle design or injector spray pattern modification would solve the problem, he spent long hours with his engineers working up additional variations to test. Dandridge trusted in objective analyses to solve engineering problems, yet the complexity and number of variables associated with combustion instability thwarted all efforts at mathematical modeling of the phenomenon.

There were other ascent engine development problems, primarily with welding and fabricating the injector, and with localized gouging of the ablative throat and nozzle in long duration firings. There were overpressure "spikes" during startup at altitude, when the engine exhausted onto a plate in close proximity, simulating the descent stage from which it would launch on the Moon. These restricted the options available to the designers for stopping instability. The Bell engine's ablative nozzle was made of thermoplastic material similar to the command module's heat shield. It resisted the extremely high temperatures (three thousand degrees Fahrenheit) inside the engine by its excellent thermal-insulation properties and by charring when overheated, leaving in place a firm, charred residual material that largely preserved the nozzle's original shape. Irregularities in the injector spray pattern could cause local hot streaks within the chamber that produced cuts and gouges in the nozzle throat, the location within the engine where the rate of heat transfer to the walls was the greatest. Baffles on the injector face improved the engine's stability, but usually made nozzle gouging and erosion worse, so the designers found themselves between the proverbial "rock and a hard place."

Two failures in the fall of 1966 gave added impetus to the quest for a solution. The first spontaneous instability, without bomb detonation in the chamber, occurred during an altitude test of the engine at White Sands; this was with a flat-faced injector. Soon afterward a baffled injector configuration failed a bomb test at Bell. More engineers, design variations, and tests were added to the schedule, but results were elusive. At one point a configuration successfully passed ten bomb tests but failed the eleventh. We had set a criteria of twelve successful tests before declaring a design stable; after that, we increased the hurdle to twenty tests.

In the midst of this activity Dandridge and Bob Thompson of Grumman Propulsion were flying from La Guardia Airport to Buffalo on one of their many visits to Bell. Seated together, they discussed engine tests and bombs intently for most of the flight. As they disembarked, four burly men in dark suits confronted them, flashed FBI shields, and hustled them into an anteroom. For the next two hours the agents questioned them severely, demanding to know what was behind all this talk of bombs that the flight attendant had overheard. It took much explaining by Dandridge and Thompson, and telephone calls to Grumman security and Bob Carbee, before the FBI was convinced that their implausible story was true.

The cut-and-try fixes dragged on without conclusive results, and by mid-1967 NASA was very concerned; they saw this problem as a potential Apollo program "show stopper." NASA hired Rocketdyne to develop an alternative injector that could be installed into the Bell engine. As soon as Rocketdyne had some injector hardware and began to obtain test results, there was a continuous series of high level reviews and visits between Rocketdyne at Canoga Park, California, and Bell at Niagara Falls, New York. General Phillips, Apollo program director, George Low, Apollo spacecraft director, and NASA propulsion expert Guy Thibideaux were involved, as was I, Joe Gavin, and Dandridge and his group for Grumman. Bill Wilson of NASA at Rocketdyne, Steve Domokos, Rocketdyne ascent engine program manager, and Dave Feld of Bell maintained close communications and a cooperative attitude, transforming what could easily have been a very sticky contractual situation into close teamwork across companies and organizations.

Rocketdyne succeeded in producing a stable injector, but the engine they designed had other problems, including hard starts, rough running and manufacturing and assembly difficulties. Bell was unable to show a positive fix on their injector despite repeated tries. At one point, in a combination of jest and frustration, Dave Feld declared, "Perhaps combustion instability is an East Coast phenomenon. Let's send our engine to Rocketdyne and have them test it out there."

To Feld's surprise, we quickly accepted his suggestion. The tests were repeated in the Rocketdyne facility at Santa Susanna, and the bomb-induced instability was still present. With the Bell engine in hand, Rocketdyne was then directed to install their injector and combustion chamber into it, retaining the Bell nozzle, valves, and mounting hardware. After some adjustments to the injector spray pattern to minimize throat erosion, this combination of the two companies' engines performed very well. Bomb-induced oscillations damped out in less that four hundred milliseconds and thrust and specific impulse[5] were within specifications.

In May 1968 Phillips and Gavin visited Bell and Rocketdyne and reviewed the latest design options and the encouraging test data. They presented three design and contracting options to Low: (1) Bell engine and Bell injector (still not positively stable), (2) Rocketdyne injectors installed in Bell engines at Rocketdyne, or (3) Rocketdyne injectors installed in Bell engines at Bell. Low chose option two, with the entire assembly put together and furnished by Rocketdyne. By June 1968 this design had passed fifty-three bomb tests without instability, and it was completely qualified in August.[6] The long ordeal of ascent engine development was over and a major item was removed from the show stoppers list. Not a moment too soon, because the heavy LM flight schedule of 1969 was almost upon us. LM had already lost its berth on Apollo 8, which was to have been the first manned LM flight with the CSM in Earth orbit. (LM-3 wasn't ready when needed to fly in late 1968.) Instead, Apollo 8 became a manned lunar orbit mission with CSM only at Christmas 1968. The original Apollo 8 mission was slipped to Apollo 9, flown in March 1969.

Stress Corrosion

The Super Weight Improvement Program resulted in the widespread use of chemical milling for structural parts on LM that rendered aluminum alloy parts more susceptible to stress corrosion, because it left a relatively rough and open-pored surface finish. Even without chem milling, certain aluminum alloys were vulnerable to this phenomenon, which in the 1960s was just being encountered and understood. Stress corrosion is intergranular corrosion occurring at the metal grain boundaries, usually visible only through a microscope looking at an etched and polished surface. A combination of steady stress and moisture or humidity is required to produce this corrosion. The stress levels can be relatively low—well below the recommended working stress of the material. Cracks and failures from stress corrosion typically do not develop until months or years after the original stress is applied—and this became a hidden time bomb for the LM structure, surfacing widely in the schedule-critical years of 1967 and 1968 and finding its way to the show stoppers list.

The most common applied stress causing this problem was "fit-up" stress, which resulted when parts did not fit or nest together exactly. When the fasteners were tightened, the parts were deflected until they made complete contact and the stress resulting from the deflection was locked into the material. Some joints, particularly the rod ends inserted into LM structural tubes, required a press fit to provide a solid, immovable connection. In such cases a predetermined stress level was purposely locked in upon assembly.

Minimizing fit-up stress required parts that fit together well or were carefully shimmed upon assembly to eliminate deflections when the fasteners were tightened. Training of assembly mechanics was revised to emphasize the importance of proper parts fit up and techniques for accomplishing it. We also reviewed the engineering tolerances on parts assemblies, particularly press fit joints.

Some stress corrosion of thin tabs was noted on LTA-1 as early as 1964, and a stress corrosion inspection and survey was performed on all LMs then under construction. The problem continued at a low level, with an occasional crack found, until a rash of cracked parts was discovered in mid-1967, beginning with LM-1. The cracks were mostly in the press fit ends of structural tubes, although some thin tabs were found cracked also. Mueller was furious. At that late date the Apollo schedule was threatened by this insidious problem—it could exist on almost any part, anywhere within the LM. The day after an inspection, a new crack could develop, because the nature of the phenomenon was chronic and progressive. LM stress corrosion was branded a show stopper.

Heavily pressured from above, I went to general quarters. Led by Bob Carbee, our best structural design and materials troubleshooters, Len Paulsrud,

Will Bischoff, and Frank Drum assembled a team from Engineering, Manufacturing, and Quality Control to inspect all the accessible structure on the LMs under construction. Visual inspection was conducted with flashlights and magnifying glasses. Suspect areas were brushed with Zy-glo dye penetrant, which glowed under ultraviolet light and enhanced the visibility of tiny cracks.

By mid-February 1968 we had thoroughly inspected six LMs (numbers 3 through 8) and over fourteen hundred accessible components. No major cracks were found. We also switched from 7075-T6 aluminum alloy to the more stress corrosion resistant—T73 temper, effective on LM-4 and subsequent vehicles. This was accomplished by retrofitting the structural tubes on these vehicles. By changing out the tubes that had been the major source of trouble and showing by inspection that fit up parts with thin tabs were not cracked, I felt we had the problem under control, and NASA agreed. By the end of the month, Mueller told NASA Administrator Webb that he was no longer worried about stress corrosion.[7]

Inspections continued throughout the LM program, and occasionally a cracked part was found and replaced. LM never suffered a structural failure of any kind, so the effect of stress corrosion on the flight missions was nil. But stress corrosion was a nagging problem that never entirely went away, as there was always a small chance that the next inspection would turn up a newly cracked part.

Battery Problems

The LM batteries were constructed with silver and zinc electrodes and liquid potassium hydroxide electrolyte and were not designed to be repeatedly charged and discharged. A limited number of recharges was allowed during ground tests, but in flight the batteries were launched fully charged and discharged until depleted. The EPS had four batteries in the descent stage, producing 3 ampere-hours per pound of battery weight, and two in the ascent stage, which yielded 2.5 ampere-hours per pound. The batteries had alternating plates of silver and zinc connected to positive and negative terminal bus bars by metallic jumpers and separated from each other by paper insulation. The plates were mounted in a row inside a vented plastic case that was filled with electrolyte. A straightforward design with seemingly not much that could go wrong. But looks can be deceiving. Before the LM program ended, in the course of resolving a long list of development problems, Grumman was forced to learn more about these batteries than even the manufacturer knew.

My confidence in the batteries waned when I visited the Eagle Picher factory near Joplin, Missouri, in 1966 as part of a tour of several LM subcontractors. Eagle Picher was primarily a manufacturer of paints and industrial chemicals based on paint pigments. They got into the battery business because some of the electrode materials, like lead and zinc, were used in their paints, and they saw a way of diversifying their product line with new applications of materials that they understood. Knowing this background did not prepare me for the sight of their factory complex—it was a sprawling industrial wasteland with many acres of low sheds and two-story brick buildings, dominated by dozens of tall smokestacks spewing great clouds of gray and white smoke. White dust covered everything—the roofs and walls of the buildings, the ground, the cars in the parking lots, everything.

The LM battery assembly area was located in one of the low corrugated metal sheds. Inside the windows were open to the swirling particle-laden atmosphere and there were layers of white dust on the windowsills and floors. At rows of worktables, strapping farm boys were doing delicate battery assembly, folding sheets of insulation over the plates, mounting the plates on the bus bars and connecting the jumpers, and installing the plate assemblies into the battery case. One muscular fellow caught my eye particularly; he was wearing a soiled undershirt and had a cigarette dangling from his lips. As I watched incredulously, the ash on his cigarette grew longer as he peered over the open case into which he had just placed a plate assembly, until it finally broke off and fell inside.

That was the last straw for me. I motioned for Eagle Picher's LM battery program manager to follow me into a glass partitioned cubicle off the assembly floor and exploded. "Don't you people have any concept of quality?" I shouted. "For God's sake, look at this place! Clean up the building, close the windows, install air conditioning and filters, prohibit smoking, make the workers wear clean clothes and smocks, maybe even replace the farm hands with nimble-fingered women."

After that visit I insisted that Grumman hold a monthly review with Eagle Picher, alternating between Joplin and Bethpage. If I could not attend the Joplin meeting, which was frequently the case, Carbee filled in for me if he could. I also prevailed upon Joe Kingfield, LM Quality Control manager, to station a resident Grumman quality inspector at Eagle Picher. It seemed like cruel and unusual punishment, to assign a Grumman man to such a place, but I considered it necessary. Our inspector filed a weekly report on Eagle Picher's activities and progress in "cleaning up their act."

All this was in the nature of a preemptive strike, since as yet no battery problems had emerged. Our concern was soon justified as erratic battery performance began to appear in tests, traced to assembly errors and contamination. As battery development progressed, many other problems surfaced. Case cracking and jumper failures during vibration tests were recurring problems, and required several modifications to design details before the batteries could pass the vibration portion of qualification testing. Underperformance was a problem that required redesign to maximize the active area of the plates and to determine the optimum electrolyte concentration. Venting of hydrogen from the case during battery operation was provided by a small plastic relief valve in the sealed cover. These valves sometimes stuck shut during cold-temperature testing, causing hydrogen gas pressure to build up and the cover to pop free of the case. In space this would cause complete battery failure, as all the electrolyte would leak out. The durability of the paper separator also gave concern. There were some short circuits between plates when pinholes developed in the separators after several ground test recharges. More durable separator material and further limitation of recharges solved this problem.

This list of miscellaneous problems continued through 1967 into 1968, giving me and Grumman a background level of concern. The batteries passed their qualification tests, although several months behind schedule, and the problems that remained did not seriously threaten the Apollo program schedule. I remained wary of the batteries until the completion of the program, preconditioned by my original view of Eagle Picher as a dirty gray scene of desolation from Dante's *Inferno*.

Tank Failures

Of the thousands of possible failures in the Apollo Mission and the LM, one of the most terrifying was rupture or explosion of a tank. Dozens of tanks on LM held the vital consumables of space exploration: rocket propellants, helium pressurant, oxygen, and water. Most of these tanks contained high levels of pressure energy and could explode in event of failure. Since all aerospace contractors used the same group of subcontractors to make their tanks and the same materials and similar processes and procedures, a tank failure anywhere in the Apollo program sent a shock wave of concern to everyone.

In mid-1965 there were two series of tank failures, both on the CSM program but directly affecting LM. CSM reaction control system fuel and oxidizer tanks each failed at their manufacturer, Bell Aerosystems, which also made very similar tanks for the LM RCS. Then a failure occurred at Beech Aircraft, which was making hydrogen and oxygen tanks for the CSM fuel cell assembly. I sent John Strakosch from LM Structural Design and Frank Drum from Materials to both companies to meet the engineers leading the failure investigations and establish an information channel to Grumman on the findings. They were also able to suggest lines of inquiry to the investigation teams.

Strakosch and Drum returned very concerned because the origin of the failures was unknown and they applied directly to LM. A list of possible causes had been prepared and was being vigorously explored by NASA, North American, Bell, and Beech. Both tanks were of highly stressed titanium and the failures seemed to originate at or near the welded circumferential seam.

Thanks to careful sleuthing, these cases were solved within a few months. The RCS tank failures were caused by the propellant manufacturer improving his process to increase the purity of the nitrogen tetroxide oxidizer that he produced. The revised production process reduced the amount of trace contaminants, including nitrous oxide, which had played a beneficial, if unsuspected, role in protecting the titanium from attack by the nitrogen tetroxide. By specifying a minimum allowable amount of nitrous oxide in the product, the problem was resolved. To assure control over the formulation of such commercial products, NASA began buying propellants for the whole program under a government specification.

The cause of the Beech problem was quite different. It occurred because a weld rod of lower strength titanium alloy had been inadvertently used to weld the oxygen tank. This was determined by metallurgical examination and analysis. A torrent of procedures and regulations followed, aimed at tightening control and accountability for weld rod and certifying that the proper alloy has been used in each weld.

In October 1966 one of the large service module propellant tanks ruptured while under pressure test at NAA in Downey. After intensive investigation the cause was determined to be incompatibility between titanium and the methanol (methyl alcohol) that was used as the pressure test liquid.[8] This oversight stimulated NASA to conduct a comprehensive survey of all the fluids to which the Apollo tanks were exposed in their lifetimes and to perform laboratory tests to establish whether the tank material and the fluid were compatible. It seemed fairly basic, but until then it had not been done. This case reminded me of an experience I had years earlier while developing a small liquid rocket at Lockheed. My test rocket failed because I had not known that the nitric acid oxidizer would attack the nickel in the high strength stainless steel alloy we were using.

Almost a year later another tank failure occurred, this time an LM descent propulsion supercritical helium tank. The inner pressure vessel ruptured while being pressure tested at the manufacturer, Airesearch. The failure originated at the weld, so at first we thought it was another case of a mislabeled weld rod. Henry Graf, supercritical helium system manager at Airesearch, impounded the remainder of the spool of weld rod used in the failed tank, and metallurgical examination showed it to be the proper alloy. Moreover, microscopic inspection of the failure showed some tiny intergranular cracks, typical of stress corrosion. Yet as far as anyone knew the tank had never been exposed to any fluids not approved for compatibility with titanium.

Henry Graf became obsessed with finding the cause of this failure. He led his engineering and quality staff through a minute examination of every step in the manufacturing process, starting with the receipt of the titanium forgings and the quality pedigree that accompanied them. At each step of the process, they looked at what had been done on the failed tank, and asked whether anything in this step was different from their process on previous tanks. Graf's careful detective work paid off, discovering a cause so trivial that a less observant investigator would surely have overlooked it. Graf noticed one minor difference in the process for this tank and those that had preceded it: instead of using new cloth pads to wipe the tank surfaces prior to welding, washed, reused cloths were employed. Examination of the washed cloths showed traces of detergent, and test samples that were wiped with them failed under combined stress and humidity testing. The trace detergent attacked titanium! There could be no more gripping example of the extreme sensitivity of highly stressed tank material and welds to contamination.

In September 1968 a descent propellant tank destined for LM-9 (Apollo 15) was found to have a cracked weld after proof test at the manufacturer, Airite Division of Sargent Industries. The test had been conducted with the tank full of distilled water at room temperature. Airite was able to verify that the proper weld rod had been used and that all fabrication processes and procedures were complied with. However, some impurities were found in the water used for the test. Frank Drum spent time with Airite in California and compared Airite's welding procedures to those of our other titanium tank suppliers, Aerojet and Airesearch. He considered Airite's procedures to be less conservative than the others in some key process steps, such as tack welding. Aerojet always did tack welding by machine inside a vacuum welding chamber, while Airite did the tack welds by hand in the factory environment before putting the tank into the chamber to complete the full continuous weld by machine.

At our insistence Airite upgraded their procedures to include the best practices of our other suppliers, and consulted with them on the "tricks of the trade" involved in making them work well. With guidance from NASA and Grumman, they also redesigned the butt weld joint to a "J-groove" design, based upon improved strength of this design in coupon samples. Airite further tightened their already strict manufacturing quality controls and improved cleanliness and housekeeping in their shops.

As part of the SWIP weight reduction effort, Grumman had prepared a lightweight descent propellant tank design and put it out for competitive bids. Airite won the competition, replacing the original descent tank supplier, Allison Division of General Motors. Allison tanks were in LMs up to the LM-5 (Apollo 11) and Airite tanks were installed in LM-6 and LM-7 when the failure occurred. LM-8's tanks had been delivered but not installed. At Grant Hedrick's suggestion, we began a cryoproof test program for the Airite descent tanks, to see if we could verify that sufficient weld safety margin existed beyond the proof pressure level to let us clear the LM-6 and LM-7 tanks without removing them. Cryoproofing consisted of testing the tank to proof pressure (one and a half times normal operating pressure) while filled with liquid nitrogen at minus three hundred degrees Fahrenheit. Welds were more brittle at this low temperature, so this was a more severe test of the fracture toughness of the weld.

An extensive fracture toughness test program was conducted on coupon samples cut from the failed tank, to show that tanks which passed cryoproofing had ample margin for all mission stresses. Samples were tested at room temperature and in liquid nitrogen, with water and with freon (which was being considered as an alternate proof test fluid), and with and without machined slots, simulating preexisting cracks or flaws.

As a result of this thorough investigation, in which materials, welding, and fracture mechanics experts from NASA-Houston and Marshall, North American, and all the tank suppliers in the program consulted or participated, we developed an acceptable lightweight tank manufactured at Airite. However, we replaced the tanks in LM-6 and LM-7 with heavier Allison tanks and installed them also into LM-8. The Airite SWIP tanks, which were also made longer to contain the added propellant required to accommodate the LM weight increase for the extended duration missions, went into LM-10 to LM-13 (Apollos 15 to 18).[9] These tanks were cryoproofed before delivery.[10]

I consulted closely with Grumman's chief technical engineer, Grant Hedrick, on these tank problems, and found him a wellspring of good advice and wise judgment. In the Airite tank problem in particular, Hedrick called all the shots. Frank Drum and his Materials Section, with their expertise in materials properties, inspection techniques, and metallurgy, were also very helpful. Throughout our tank problems, we worked closely with Joseph Kotanchik, branch chief of Structures and Mechanics at NASA-Houston, and his very capable group. They made the full facilities and expertise of NASA available to us and provided valuable suggestions. At one point during the Airite tank weld fracture toughness test program, all ten of NASA-Houston's fatigue test machines were loaded with our coupon test specimens.

We were nervous enough about the LM tanks that we sought advanced test and inspection techniques to further verify their quality, as in our cryoproofing program to recover from the Airite tank failure. The routine inspections and tests that we required on all tanks included weld X-ray and dye penetrant (Zy-glo) inspections and proof pressure test with water at ambient temperature. The supercritical helium tank was also cryoproof pressure tested at its extremely low (cryogenic) operating temperature of –450 degrees Fahrenheit.

We experimented with installing a network of acoustic sensors on the ascent propellant tanks during proof pressure tests. Pinging sounds were emitted as the tank material stretched under pressure, and the system could triangulate with multiple sensors to locate the point from which the sound originated. Dye penetrant inspections were performed at these points of origination to see if incipient flaws could be detected. Results were inconclusive and no flaws were detected, so we abandoned this approach.

Like most LM components, the tanks were extremely sensitive to the slightest mistake at any point in their fabrication, assembly and test process.

This meant that they were totally dependant upon the skill, craftsmanship, and integrity of their builders, as was the whole LM and all its parts. A prime example of a tank builder's integrity and concern was shown by Bruce Baird, ascent propellant tank manager for Aerojet General at Downey, California. Baird, a short man with a serious but youthful face, came into my office one day and declared he had a problem. He showed me color photos of a lunar module ascent tank that had recently completed acceptance tests at Downey and was ready for delivery. The problem was that, in the final heat treat, this tank had turned a dark purplish color, whereas all other ascent tanks were colored light straw after heat treat. Something different had been present in the atmosphere of the heat treat furnace for this tank, and Baird did not know what it was. Although the tank had passed its proof pressure test, he feared that it might have been damaged in some way that would reveal itself later, perhaps in flight.

Hedrick and our other experts went out to Downey to see the suspect tank, and examined its records and the whole Aerojet tank fabrication and test process. They concluded that the tank was probably all right, but since the discoloration could not be explained and Baird felt so strongly about it, they agreed not to accept this tank for flight but to use it for the ultimate test-to-failure demonstration in the tank qualification test program. In the test to failure the purple tank exceeded its design ultimate pressure, so there had been nothing wrong with it after all. Nonetheless, we admired Baird for his dedication to quality and the success of the Apollo mission. LM never had a tank failure or leak in flight, thanks to Graf, Baird, and many others like them at Grumman and our suppliers.

10

Schedule and Cost Pressures

In those heady days at the beginning of Project Apollo, cost was never mentioned, except when NASA questioned whether we had enough money to do what was necessary. We were free to plan, design, expand staff and facilities, and do whatever it took to get the program moving. Congress had almost unanimously approved President Kennedy's goal of landing men on the Moon by the end of the 1960s, despite its uncertain price tag.[1] The initial program estimate was $8 billion, but it was commonly expected to cost $20 billion. Congress funded all federal expenditures yearly, so even a long-duration program such as Apollo got its money one year at a time, with no guarantee of how much money, if any, would be approved the following year.

Before long NASA, just like all other agencies and programs, had to fight for money in Congress. Congress began questioning and trimming Apollo's costs in 1965, when NASA's budget exceeded $5 billion. The rapid growth of Apollo expenditures brought charges of waste and lack of control. NASA spent $5.1 billion in fiscal year 1965 and $5.2 billion in fiscal year 1966, of which $2.5 billion in fiscal year 1965 and more than $3 billion in fiscal year 1966 was for Apollo. The NASA fiscal year 1967 budget request of $5.58 billion was cut to $5.012 billion by President Johnson and further trimmed to $4.968 billion by Congress. Apollo's budget escaped unscathed, but the follow-on Apollo applications program was all but deleted.[2]

The LM program at Grumman directly felt this budgetary belt tightening. Grumman's expenditures had risen dramatically, from $135 million in fiscal year 1964 to $350 million in fiscal year 1966, attracting the attention of NASA management. NASA's first move against Grumman's increased spending fit well with a desire that had been building in Headquarters to convert the Apollo contractors to incentive contracts from cost plus fixed fee. With an incentive contract the contractor's fee was determined by program accomplishments as well as cost and schedule performance. Both Administrator

James Webb and George Mueller believed this to be necessary, not only to motivate their contractors financially to improve performance but also to show Congress that they were being hard-nosed in managing them. The Department of Defense under Secretary Robert McNamara was also moving to incentive contracts.

Incentive Contract Negotiation

In March 1965 Joe Shea kicked off a major exercise aimed at renegotiating Grumman into an incentive contract. The effort began with NASA examining the updated material we had prepared, which gave the detailed work statements, schedules, and estimated cost to completion for each contract line item. The top level LM system specification, the overall document describing the LM and its systems, and the performance and interface (P&I) specification, the top-level summary of LM's technical performance and characteristics and its technical interactions with any other element of the Apollo-Saturn system, were also prepared for negotiation.

A large, high-ranking NASA contingent arrived in Bethpage to conduct the review and negotiations, led by Joe Shea, R. Wayne Young (who had replaced Bill Rector as the LM project officer), and Tom Markley from the Apollo Spacecraft Program Office. Joe Gavin and Bob Mullaney headed the Grumman effort. Rathke and I led the large Engineering supporting effort, which included assistant project engineers Carbee, Whitaker, and Coursen, the System and Subsystem engineers and section heads, the cognizant engineers, and any other support they required. As in the original contract negotiations in Houston, our people paired off with their counterparts into fact-finding and negotiation teams, meeting in the many conference rooms in Plants 25 and 5.

Rathke and I worked all day on the many other engineering efforts in progress: finishing the design, getting the drawings out, trimming LM weight growth, and resolving technical problems. At 5:00 P.M. we assembled the team leaders and reviewed the progress and status on each line item and work package. Often these meetings lasted until 9:00 or 10:00 at night. After two weeks of fact finding, NASA started making their recommendations.

NASA was looking for 25 to 40 percent cuts in cost and manpower. We were shocked; this certainly was nothing like the original negotiations. None of us believed we could do the job for the cost NASA wanted, and I set the team leaders off to prepare rebuttals. For the next several days we worked with the team leaders to refine their rebuttal evidence and presentations. I thought we had strong positions in every area.

After the rebuttals were made to the NASA teams, our team leaders reported that most of their NASA counterparts agreed with them, with usually minor exceptions. The total estimated cost was going up, even beyond our initial position, because of oversights and underestimates that became apparent during the detailed negotiations. Joe Shea became increasingly testy; he was gruff with us in our management interface meetings, and there were reports that he had been dressing down his own troops for failing to reduce LM costs.

After a month's effort, Shea abruptly canceled the whole exercise. Most of the NASA delegation returned to Houston, but Young, Bill Lee, and a few others remained to tie up loose ends. NASA agreed to use the unofficial positions agreed by the team leaders in each line item and work package as the planning baseline for the LM program.

In June 1965 Bob Gilruth met with Grumman president Clint Towl to review the outcome of the Shea exercise. They agreed that they could not baseline an incentive contract yet. Gilruth then informed Towl that NASA was imposing a lunar module management plan, limiting Grumman to spend no more than $78 million for the last quarter of fiscal year 1965, an amount considerably less than Grumman's estimate.

We worked with our subcontractors to negotiate incentive contracts with them. Our subsystem engineers, cognizant engineers, and the subcontracts managers led this effort, which continued throughout the summer. We also reviewed and reestimated the Grumman in-house work packages. In September Grumman submitted an incentive contract conversion proposal to NASA that became the basis for continuing negotiations during the fall and winter. Engineering support to LM program management was continually provided as required, but it was a "low-key" activity carried out by small teams, often over the phone. NASA's emphasis had changed from cutting the estimates to trying to understand the most likely cost. In this improved environment, agreements came quickly at the team level and were supported by program management.

In February 1966 agreement was announced on an updated contract for the Grumman LM program containing incentives on schedule, cost, technical and mission performance. It carried the LM program through 1969 at an estimated cost of $1.42 billion. The contract also invoked a revised set of LM specifications and statement of work, which defined Grumman's responsibilities for the remainder of the program. Gavin and Mullaney held internal meetings to publicize the incentive provisions of the contract, so that everyone working on LM would know what the specific short term goals and priorities were. Highest priority was meeting the scheduled ship date for LM-1, the first flight LM, of 15 November 1966.

Hjornevik Review

Unfortunately, reaching agreement with NASA and our subcontractors on schedule and cost goals was no guarantee of achieving them. The ink on the new contract was barely dry when our forecasts at Bob Mullaney's weekly program meeting showed further cost growths and schedule slips. This continued into the spring, despite vigorous efforts by the subsystem engineers and subcontract managers to contain it. The deteriorating situation attracted the attention of Grumman's newly appointed president, Lew Evans.

Evans was short and stocky, with a full face, black hair, twinkling eyes, and mischievous smile. He radiated energy and charisma, the natural leadership ability that sweeps others willingly into its train. When talking to Evans you felt that his whole attention was focused on you and your concerns as his clear blue eyes bore into you with total concentration. Even his many facial tics (eye blinking, mouth stretching, neck rotation) enhanced the impression of a human dynamo bursting with energy. He genuinely liked people and was interested in them as human beings, not just cogs in his enterprise. Above all he conveyed a spirit of optimism and unlimited opportunities for those who followed him, and a belief that the future for Grumman and its people was boundless.

Evans was born in what is now North Korea, the son of a globe-trotting mining engineer. He lived in many places, including several years in Mexico, where he became fluent in Spanish and developed a taste for spicy food. After graduating from the University of California in 1942 he served with distinction in the U.S. Army Air Corps. He graduated from Harvard Law School in 1947, was admitted to the bar, and served four years as assistant counsel of the Bureau of Aeronautics in Washington, D.C. While at the bureau he observed the key roles that politics and personal relationships played in obtaining contracts from the navy, or any government agency.

Evans joined Grumman in 1951, his interest in marketing and customer relations leading him to become vice president and director of Business Development in 1960, heading a small team of marketing professionals. His department was a departure from Grumman's usual technically dominated approach. Staffed mostly with retired military officers, lawyers, and salesmen, Business Development's focus was capturing new business and keeping existing customers satisfied. Saul Ferdman, one of the few engineers on Evans's team, was thoroughly imbued with his "keep the customer happy" philosophy.[3]

Evans recognized that Grumman had cultural differences with NASA and a serious problem in making NASA's leadership believe that the LM program was not playing second fiddle to the navy within Grumman. When he became the company's president, he was anxious to do something about this situation. The opportunity came very quickly.

NASA's Bob Gilruth called Evans to congratulate him on his elevation to president and to request a meeting with him to review Grumman's schedule and cost performance difficulties on LM. Gilruth and Joe Shea met Evans in Bethpage and demanded that he control the situation. Evans summarized the actions Grumman had already taken and invited NASA to provide experts to review Grumman's management of the LM program. He said he would regard the NASA reviewers as his "personal management analysis staff."

Wesley L. Hjornevik of the ASPO, assisted by R. Wayne Young, LM project officer, and Thomas Markley, CSM project officer, led a large NASA review team drawn from Houston and Headquarters. They moved into Plant 25 in Bethpage, and some of the team members also visited Grumman's major subcontractors. They attended some of our ongoing meetings, such as Gavin and Mullaney's weekly program meeting, Rathke's and my Engineering project meeting, and Coursen's ground-support equipment meeting. They studied the cost and schedule performance of Grumman and its subcontractors, comparing predictions and promises versus actual results. The review team interviewed Grumman's corporate and LM project management, and examined our organization, staffing, procedures, and delegation of authority and responsibility. It was an intensive, highly professional review, and after ten days the Hjornevik team reported its findings to NASA and Grumman top management.

The report was critical of Grumman's management of the LM program, but it also contained constructive suggestions for improvement. The review team noted that recent cost growth and schedule slippages had been concentrated in the subcontractors, rather than the in-house Grumman effort. They attributed this to the absence of a focal point for subcontractor management in Grumman's LM program organization, and to splintered responsibility for dealing with subcontractors between Engineering and the Business Office. Hjornevik and Shea also felt that Grumman had not fully implemented work package management or related the defined work packages to the program's leadership. Although work packages had been prepared to define, estimate costs, and schedule every part of the program, the work package managers had little authority and often competed with one another for resources.

The team considered that GSE was still seriously underestimated and undersupported at Grumman, and that although GSE Engineering's performance had improved, with about 80 percent of the designs either released or on schedule, the problems had shifted to GSE Manufacturing, where schedule performance was still terrible. Hjornevik's team was unable to find a coordinated overall plan for GSE. They urged Grumman to demand better GSE performance by holding daily program management meetings and consolidating the GSE Manufacturing facilities, which were scattered throughout the Bethpage complex. NASA also recommended that Grumman off-load its internal GSE workload by purchasing GSE end items from companies such as General Electric, North American, and McDonnell, which were already producing similar GSE units for the Apollo and Gemini programs that could be redesigned or modified for LM.

Lew Evans reacted swiftly and decisively to the Hjornevik Team's findings and recommendations. Senior Vice President George Titterton moved into an office next to Joe Gavin's in Plant 25, eschewing his far more luxurious quarters in Plant 5's "mahogany row." Evans put him in charge of all space programs at Grumman, with at least 80 percent of his effort to be spent on LM. Although he was not relieved of his other corporate responsibilities for program management, business, and contracts management, Titterton increased his already heavy concentration on LM to almost full time. I marveled at how Joe Gavin continued to perform in his usual competent, unruffled manner, despite Titterton's seeming incursion into his authority.

Many other top level LM program organization and key personnel changes took place at Evans's direction.[4] Bob Mullaney was made Titterton's assistant and would be reassigned to Lew Evans's staff when LM's management performance improved. To me it seemed that, as the program manager, Mullaney had been made the scapegoat for Grumman's poor cost and schedule performance on LM. As in professional sports, if the team loses the manager often gets fired. In addition, Joe Shea and Mullaney did not get along well. Mullaney could not resist baiting Shea with provocative comments, sometimes in public. Shea was still smarting from the standoff with Grumman in his incentive contract conversion exercise where Mullaney outmaneuvered him to a draw.

Bill Rathke replaced Mullaney as LM program manager and I succeeded him as LM Engineering manager. My job of LM project engineer was not filled, but Carbee, Whitaker, and Coursen were made Subsystem, System, and GSE project engineers, respectively. I viewed this with chagrin—I lost Rathke's steady guidance and close involvement with LM Engineering and had to assume much of his workload.

Brian Evans became LM subcontracts manager, reporting to Gavin. He was responsible for the performance of Grumman's LM subcontractors and for assuring that our subs received adequate support from wherever it was needed within the LM program. The individual subcontract managers reported to Brian Evans and their authority was strengthened, making them the senior management interface with the subcontractors. The subsystem and cognizant engineers reported to them on any matters affecting the subcontractor. Each subcontract manager led a combined Subcontracts-Engineering team with support from other LM departments as required. This provided the focused subcontract management approach that the Hjornevik team had recommended so strongly. Brian Evans brought in several senior procurement managers to strengthen the ranks of subcontract management and set high standards for them to meet with their subcontractors. Although mild mannered and even self-effacing, he created an environment in his organization in which his subcontract managers did not hesitate to demand a hearing from the president or CEO of their subcontractor, if they thought the problem warranted it.

In the critical area of GSE, Daniel Culleton, an experienced plant manager from Grumman Manufacturing, was made GSE Manufacturing manager, reporting to Rathke and Gavin. He was also given more resources. Grumman purchased a fifty-thousand-square-foot building in Syosset and refurbished it for efficient GSE fabrication and assembly. GSE Manufacturing was consolidated at this one location. GSE budgets and staffing were again increased and support from Ed Dalva's Corporate Integrated Logistics Support Department was strengthened.

To complement Grumman's actions NASA appointed a management review team headed by Wayne Young to meet with Grumman monthly to assess status, tackle problems, and follow up on the effectiveness of the changes made in response to the Hjornevik report. NASA appeared pleased with Lew Evans's prompt and stern reaction to their review and the concentrated attention Grumman's senior corporate management was devoting to the LM program. They felt that at last Grumman was going beyond its traditional dedication to the U.S. Navy and was considering NASA and its Apollo program a very high priority customer.

Within three months after this management shakeup, Grumman's performance showed some improvement. Subcontractor performance generally improved, showing less schedule slippage and cost growth. Although better, these problems were not eliminated, and management pressure and scrutiny intensified.

Grumman gradually turned the corner on its GSE problems during the summer of 1966. Schedule slippages and parts shortages decreased significantly as the new GSE factory became operational and the added people on the program were trained and became effective. The workload on GSE Engineering and Manufacturing was reduced by procuring many end items through competitive bidding, and buying and modifying "common use" GSE from North American and from the Gemini program. Some GSE items were furnished by NASA without charge as surplus inventory, in which case we saved money as well as time and manpower. By the end of the year, GSE availability was no longer constraining the schedule along the critical path[5] of the LM program.

Schedule Iterations

In July 1966 LM-1 through LM-4, all the flight LMs in the factory, were late to schedule and still slipping. LM-1, the first flight LM, was delivered to KSC on 22 June 1967, five months late to the prediction Shea had made to Congress in October 1966.[6] The Apollo 8 mission orbited the Moon without LM. It had originally been scheduled as the first manned LM mission, to be flown in Earth orbit with the CSM. When LM-3, the first manned LM, was unable to meet the Apollo 8 schedule, NASA did not want to lose Apollo momentum after the successful manned CSM Earth-orbit flight of Apollo 7. At George

Low's urging, they rapidly developed the spectacularly successful CSM lunar-orbit mission, which thrilled the world on Christmas Eve 1968.

Larry Moran and his Scheduling Group struggled valiantly to keep up with the endless schedule iterations and to provide LM management with a broader view of whether we were gaining or losing in the battle to maintain schedules. They collected PERT input data daily from every corner of the LM program, feeding it into the computer overnight and updating the wall displays of schedules and networks early in the morning. Moran drove himself and his group relentlessly, even as they tried to instill schedule discipline into the many diverse groups from whom they collected data. Throughout all the long hours and iterations, Moran maintained unfailing good humor. I remember coming across him and several of his cohorts late one evening in Plant 25. I stayed while Larry showed me the latest problem areas in the schedule and led me through the critical path on the network diagram. One of his men said he wished he were home, to which Larry replied (in jest, but with a straight face), "Whaddaya mean you wish you were home? You *are* home!"

I believe that Larry Moran's dedication to the LM program ultimately cost him his life. Smoking heavily and living on coffee and junk food while keeping long hours on the job, he was physically run down. On Memorial Day weekend 1967 he fell ill and died from a fast-moving, unidentified infection. I remember him with warmth and admiration as one of the unsung heroes of the LM program.

Subcontractors and Suppliers

Although we at Grumman were fiercely proud of *our* LM, we recognized that we had to share ownership of this marvelous machine with NASA, who directed, funded, and contributed heavily to its technical development, and with Grumman's network of subcontractors and suppliers. About 50 percent of the work on LM (by dollar value) was done for Grumman by hundreds of companies located in forty-six states. Ranging from giant corporations such as RCA and United Aircraft to small specialty shops making unique components and parts, these companies had thousands of people designing and building LM subsystems, components, and materials. Their dedication and expertise was vital to the LM program. Our vendors excelled in advanced, highly technical specialties and they had a long history of supplying similar flight equipment for aircraft, which gave them valuable design and manufacturing know-how. Without Grumman's capable subcontractors and suppliers the LM would never have reached the launch pad, much less the Moon.

I came to know many of these companies, their people and their products, as the LM program moved forward. I regularly attended quarterly reviews and other important meetings with our major subcontractors: RCA, STL,

Bell, Hamilton Standard, Rocketdyne, and Marquardt. When problems arose, my involvement with them increased, as with, for example, the previously described problems with rocket engines, tanks, and batteries. I also visited many of our component suppliers, usually because they were having problems, but sometimes just to see what Grumman could do to help them meet their schedules and performance requirements. In this latter mode I made two West Coast vendor trips in 1966 that acquainted me with the key people, manufacturing processes and technical issues involved with a number of essential LM components.

Grumman's subcontractors and suppliers actively supported the LM flight missions by standing ready to provide whatever assistance we might require. The major subcontractors had technical representatives with us in Houston during the missions, sometimes in the mission evaluation room (MER) in Building 45. All suppliers were on call from Grumman at Houston or the mission support room in Bethpage (MSR-B). On several occasions they tracked down test and inspection records providing vital information about the specific components involved or suspected in flight anomalies. An "anomaly" was any performance or event noted in flight that was different from "nominal" (normal or expected). In some cases suppliers ran specially requested tests on their units in their laboratories to check performance in a unique situation encountered or expected in flight, as the flight mission was underway.

Through subcontractor and supplier visits plus the major subcontractors' quarterly reviews and special meetings as required for problem solving, I maintained close contact with the subcontractors, their key people and their principal design, manufacturing and test problems. This enabled me to make informed judgments to help them, balancing test requirements versus schedule and authorizing performance deviations or parts substitutions. Subcontractor contacts later proved invaluable in flight mission support. I could discuss in-flight problems comfortably with their people and could clearly visualize the hardware and the critical manufacturing and test processes.

As the Apollo program moved ahead I became increasingly proud of our subcontractor and supplier team. The breadth of talent they represented helped us and NASA through the rough spots on the flight missions, including the LM lifeboat rescue on Apollo 13. It was a privilege to work with such fine people who were self motivated far beyond the scope of their obligations as equipment suppliers—they shared with us and NASA the dream of reaching for the Moon.

On Schedule at Last

By the fall of 1968 Grumman was essentially on schedule with LM deliveries, placing us in good position to support the scheduled launches of Apollos 9,

10, and 11 in March, May, and July 1969. If all went well Apollo 11 might be the first landing on the Moon. We still could not ease up, however: there were too many scheduled PERT events to go, any one of which could be hung up or changed by unforeseen events. But at least we could glimpse, in the far-off and hazy distance, the tantalizing goal of our journey. The Moon was phasing into a destination with a schedule.

11

Tragedy Strikes Apollo

It was a Friday evening in midwinter, 27 January 1967 to be exact. I left work early, at a little after six o'clock. It was my son Edward's eighth birthday, and I wanted to get home in time for dinner with the family, to sing "Happy Birthday" to Ed as he blew out the candles. I turned on the car radio to the news and listened glumly as the reporter described the escalating fighting in Vietnam. The familiar road wound ahead in the darkness through the bare woods and farm fields approaching Huntington. A special bulletin snapped me out of the quiet reverie of an oft-repeated drive: "NASA reports that there has been a fire in an Apollo spacecraft under test at Kennedy Space Center!"

No further details were forthcoming, but when I reached home my family was in front of the television and the children turned to me, the smaller ones shouting excitedly, "Daddy, there's been a fire, and the astronauts were killed!"

Poor Edward had to share his birthday attention with the TV while Joan and I kept one ear cocked for further details. As I watched our happy little boy smilingly devour a large slice of birthday cake, more grim accounts filtered through. Three astronauts were dead: Gus Grissom, Ed White, and Roger Chaffee, the prime crew for the first manned Apollo mission. Spacecraft 012 had been scheduled for launch into Earth orbit atop Saturn 1B 204 on 21 February 1967. The fire had occurred during a practice launch countdown at Launch Pad 34, with the crew inside the Apollo command module, mounted with its service module on the huge, but unfueled, Saturn 1B booster rocket. Few other details were available.

The next morning I drove into work in the brilliant cold winter sunshine for an eight o'clock meeting with Tom Barnes to review advanced mission planning options. NASA was already thinking of upgrading the last few LMs for more ambitious lunar exploration, if the Saturn booster payload could be increased to permit LM to grow heavier. Once there it was hard to concentrate on advanced missions with the previous evening's disaster still unfolding as we spoke. From Herb Grossman, Grumman's Engineering manager at KSC, I learned further disturbing details. The fire had been very hot and fast moving. Pressure buildup within the command module burst its crew compartment structure open within thirty seconds after the first alarm from the crew, leaving the interior a charred ruin. The ground crew in the white room adjacent to the command module was initially blown back by the fireball emitted when the cabin burst, and lacking protective equipment, were retarded by the intense heat and smoke in their efforts to remove the boost protective cover hatch and open the outer ablative hatch and the inner metal hatch, which opened inward and were each secured by several mechanical latches. Some of the ground crew watched in horror as the surveillance TV camera trained on the spacecraft briefly showed Grissom and White at the window, futilely fumbling with the hatch while outlined in an ominous orange glow of flame. It took five and a half minutes after the alarm to get the hatches open, and by then nothing could be seen inside except impenetrable black smoke. The crew never had a chance to escape, but they had the awful knowledge of what was happening to them.[1]

Gradually we realized that what had occurred was not only a personal tragedy to three astronauts and their families and friends but also a major setback to the Apollo program. The cause of the fire had to be isolated and removed, but beyond that, all other fire hazards must be identified and eliminated anywhere in the CM or LM. I was alarmed at the reports about laboratory tests done by NASA and the air force that showed extreme flammability of many materials in pure oxygen, even if they burned slowly in air. I asked Bob Carbee to assign our Materials engineers to gather the available information on this phenomenon.

Even worse, as I talked with Carbee and Barnes in the half-empty expanse of Plant 35 (many engineers had been given the weekend off), we worried about how we could all have been so blind to an ancient hazard, which in retrospect was blatantly obvious. If we were so obtuse about fire, how many other serious hazards had also escaped our faulty vision? At the very least a total, searching review of LM hazards and protective features would be required, with additional new eyes added to those of us who had ceased to see the apparent.

As at a wake, we shared stories about the deceased. I had only met Grissom briefly at the M-1 mockup review, but I had talked to White and Chaffee a few times when they were working on lunar egress during and after the M-5 review. John Rigsby, Gene Harms, and Howard Sherman had worked very closely with White on the TM-1 with the Peter Pan rig, developing improved versions of the forward hatch, ladder, and descent-stage lunar experiment bay. We were all sobered and saddened by this grim turn of fate, made especially painful by the feeling that someone, somewhere in the vast Apollo program should have recognized the fire hazard and spoken out about it. And that someone could have been me. (In fact, Hilliard Paige, general manager of General Electric's Apollo Support Division, had sent ASPO manager Joe Shea a letter in September 1966, pointedly warning of the danger of fire during ground tests in pure oxygen and urging that action be taken to reduce the amount of flammable material in the crew cabin.)

It was little comfort to rationalize, as we briefly tried to do, that in a program as huge and complex as Apollo something was bound to go wrong. Or that three astronauts had already been killed in the line of duty, in crashes of their T-38 jet trainers.[2] There was no way to avoid the realization that the Apollo program was in crisis and we were going to have to work very hard to dig our way out of it.

Reaction and Redesign

NASA and Congress each conducted investigations of the cause of the Apollo 1 fire and the recommended corrective actions.[3] Prior to the fire, the U.S. manned spaceflight program had relied upon designing to eliminate potential ignition sources as the principal way to prevent fires. If a fire occurred in the crew cabin in space, it could be quickly extinguished by dumping cabin pressure, provided the crew were in their spacesuits. No special provisions were made for extinguishing cabin fires on the ground, nor was major effort made to minimize the amount of flammable material contained in the crew compartment.

The investigations concluded that the Apollo 1 fire had most likely been started by an electrical spark, probably in or near the environmental control system module, that ignited wire insulation. It quickly spread to the highly flammable nylon, including the ubiquitous Raschel netting used to stow checklists, flight plans, and other materials used by the crew, and then raced throughout the cabin. Aluminum lines containing flammable water-glycol coolant melted in the heat and sprayed more fuel into the fire. The plastics burned and gave off toxic gases and dense smoke, which asphyxiated the astronauts once their spacesuits ruptured. (They were not incinerated, as early reports had claimed.) With the cumbersome arrangement of a boost protective cover hatch and two inward opening spacecraft hatches, the crew's doom was sealed. The white room support crew had no awareness of, or training and equipment for, fighting fire in the cabin, should it erupt.

Recommendations included: minimizing flammable materials in the cabin, protecting and providing fire breaks for any flammable material that remained, improving the quality of spacecraft wiring and plumbing, using stainless steel lines to carry water-glycol coolant, and studying two gas oxygen-nitrogen cabin atmospheres. These could apply to both CM and LM. Then uniquely for the command module: provide a single quick release outward opening hatch, consider two gas atmosphere and fire-extinguishing systems for ground test, and train and equip the white room team for fire fighting. In addition the investigating board demanded correction of the conditions that caused the many deficiencies they found in command module design and engineering, manufacture, and quality control.

Long before the formal investigations were complete, Grumman embarked on a thorough inventory of all materials in the cabin, characterizing, by test if necessary, their flammability in the LM cabin environment of 5 psia pure oxygen. (LM ground tests and checkout were run with ambient air in the cabin.) NASA enlisted us in a major review, covering all aspects of safety hazards and spacecraft quality, with an emphasis on eliminating flammable materials, potential ignition sources, and quality defects. Spacecraft wiring and water-glycol coolant lines received special attention. NASA sent materials expert Robert L. "Bob" Johnston to Bethpage for several weeks to work with our Materials Group leadership organizing the materials characterization program and developing new design guidelines.

They banished nylon from the LM cabin, along with some forms of fiberglass cloth. It was replaced by Beta cloth, newly developed by Corning Glass. Beta cloth was nonflammable and nontoxic, but it had poor wear resistance and was prone to flaking. Velcro was largely replaced by metal snap fasteners, grommets, and Beta cloth ties. Other plastics, particularly polycarbonates, which gave off toxic fumes when burned, were replaced with sheet metal where possible. Kevlar insulation was used on electrical wiring; it was fire retardant—charring in flame but smoldering or going out when flame was removed.

New design guidelines for electrical wiring were adopted. First, wire bundles and connections were to be neatly combed and rigidly supported with clamps or Beta cloth ties at least every four inches. No "rats' nests" (uncombed jumbles of wiring) were allowed at junction boxes, splices, and connectors. Second, fire-retardant potting (newly developed) was to be used on electrical connectors and switches and X-rayed after the potting cured. (When cured with a heat lamp, potting became firm but slightly flexible and adhered to the wires' insulation and the connector, protecting the connector from moisture.) "Birdcaging" (wires pushed into an arc shape, rather than straight) was not allowed. Third, circuit breakers and some switches were to be covered on their back sides with hand-tied Beta cloth "booties," providing fire protection to the plastic in the unit's body or innards and preventing short circuits by floating metallic objects (such as screws, washers, etc.) in zero gravity.

In the ECS system, aluminum tubing in the LM cabin was changed to stainless steel and rerouted to shield it from accidental damage by the crew. We also increased the flow rate of the LM cabin dump valve to speed fire extinguishing if needed in space. We participated in a joint group with NASA and North American, which reexamined the cabin atmosphere issue for CM and LM. For the command module a change proposed by Max Faget was adopted: For ground tests requiring spacesuited astronauts, the CM would be pressurized to 16.7 pounds per square inch absolute (psia)[4] with 60 percent oxygen (O_2)/40 percent nitrogen (N_2), instead of pure O_2 as before. After launch the cabin pressure would bleed down to 5 psia using pure oxygen for the spacesuit loop and for cabin leakage makeup.

Since LM only operated in space and was unmanned at launch, it could be kept at the 5 psia pure oxygen originally chosen. At launch, when LM was unmanned inside the spacecraft/LM adapter with all systems turned off, it had been planned to pressurize the LM cabin with 16.7 psia pure oxygen. (For both CM and LM, the 2 psi differential above ambient was used to keep humidity and contaminants from getting into the cabin.) In the postfire scrutiny, Marshall decided that this was unacceptable because oxygen leaking or venting from LM could combine with hydrogen leaking from the SIVB stage just below it, possibly causing a flammable or explosive mixture inside the SLA. To reduce this hazard, the LM cabin at launch was instead charged with 20 percent oxygen–80 percent nitrogen, which was bled down to 5 psia in space, with pure oxygen makeup. During a lunar mission the LM cabin was vented to permit opening the front hatch. Upon the first repressurization the cabin would contain 5 psia pure oxygen.

By the time implementation of these changes was in full swing, I had left LM Engineering to lead LM Spacecraft Assembly and Test. The engineering redesign effort was led by John Coursen, who succeeded me as LM Engineering director, and his deputy, Erick Stern. Sal Salina was in charge of the flammability test program, with major assistance from the Structural Design and the Materials Sections. While under heavy pressure from upper management to minimize the schedule slippage caused by the flammability changes, Coursen and Stern implemented them completely and efficiently. They deserve great credit for leading LM Engineering's recovery from a dark hour.

About two weeks after the fire, and shortly after I transferred into S/CAT and moved into the temporary office trailer complex behind Plant 5, we were visited by a top Apollo management delegation from NASA led by George Mueller, Gen. Samuel Phillips, and Joseph Shea. After meeting with Lew Evans, Titterton, Gavin, and other executives, they spent the day holding meetings with large groups of Grumman LM people throughout Bethpage, assuring them that the Apollo program would continue despite the setback of Apollo 1, and that it was even more important for everyone to do his job correctly and efficiently. Schedule was important, but quality was even more so. "Do it right the first time" was the slogan of the day. Apollo's salvation lay in the skill and dedication to quality of its people.

In S/CAT we held the mass meeting on the floor of the LM Assembly area in Plant 5, a very large, high-ceilinged clean room, totally white in walls, ceiling, and floors. Everyone wore the required white smocks, caps, and cloth booties, including the speakers: Mueller, Shea, Evans, and Gavin. They stood together on a raised work platform where they could look out over the uplifted sea of white-clad faces. It was like a revival meeting, with Mueller and Evans the most inspiring, asking all of us to dedicate our efforts to the memory of the lost astronauts and assuring that their deaths were not in vain. I think we all felt reassured that the program and company leaders had the will and the vision to pull us out of the problems.

After the big meeting in S/CAT, I led the NASA executives on a tour of the LM's on the assembly floor and also to adjacent shop areas in Plant 5 where subassemblies were prepared for installation. On the assembly floor they scrutinized the LM's wiring and plumbing. Although generally pleased with what they saw, as it apparently looked better than the workmanship in the command module, they insisted that we reexamine every detail of these installations. When we entered the subassembly areas in the shops Mueller, a nonsmoker, suddenly produced a cigarette lighter, which he proceeded to use to test the flammability of many components that he saw being assembled. I blanched as he pounced upon wire bundles, switches, and circuit breakers and immersed them into his flame, staring quizzically at the result through his thick horn rimmed glasses. In the panel shop, where the control and display panels for the LM flight stations were assembled, I thought the foreman would have a heart attack when he saw Mueller whip out his lighter and hold the flame onto wiring and potting in the back of a newly assembled flight instrument panel. The only substance that burned in these forays was the potting, which continued to support a candle-like flame after the lighter was removed. Everything else just charred and smoldered when the ignition flame was taken away. (Fortunately Mueller did not come across any of the nylon netting, which would have given him quite a show.) I'm not sure what these ad hoc tests proved, since ambient air was a less severe environment than the 5 psia pure oxygen in which LM operated in space, but they seemed to satisfy Mueller's curiosity and the damage that they caused was not major. The panels and assemblies would all require rework anyway when the material substitutions and other fire related changes were finalized.

We adjourned to our spartan conference room in the trailers, an unimpressive room in faux wood paneling and beige asphalt tile floor, with a low ceiling inset with fluorescent lights. We gathered around the two plastic-topped metal tables at the front of the room; some sat in the front row of the hard plastic chairs with schoolroom writing arms that largely filled the room. The NASA men shared with us their first impressions of what the corrective actions would entail, emphasizing the need to remove flammables and potential ignition sources from the LM cabin, and to rededicate ourselves to quality in every detail. They impressed upon us the need for Grumman in particular to minimize the schedule impact of the changes because we were so far behind the rest of the program to begin with.

During these discussions I first noticed indications that the tragedy had struck Joe Shea very hard. He was somber and muted, without his usual flashes of wit and puns, and lacking the self-assurance and confidence that were his hallmarks. When I had a few words with him privately, he said he would never forgive himself for being so blind to the danger. He told me how close he had come to being inside Spacecraft 012 during the test, and mused whether that might have been better.[5] I urged him not to blame himself; we all shared the blame of overlooking the obvious.

In the following weeks, Shea publicly and personally accepted the blame for the Apollo 1 accident and was perceived by the NASA leadership as increasingly showing signs of stress, despite his vigorous efforts to define and organize the recovery activities. Gilruth and other NASA leaders became concerned that Shea's worsening frame of mind might affect his judgment on program matters and concluded it was not fair to Shea to keep him in such a demanding job. In early April, convinced by earnest entreaties and blandishments by Administrator Webb, Deputy Administrator Seamans, and Mueller, Shea agreed to relinquish his position as ASPO manager and become Mueller's deputy at NASA Headquarters in Washington. Once there, he found he had little real work or responsibility, and in July 1967 he resigned from NASA to become engineering director of Polaroid Corporation, located near Boston.[6] Joe Shea quietly left the grand stage of Apollo, leaving a legacy of monumental contributions to the definition, organization and implementation of the program. He provided objective analysis, sophisticated engineering judgment and practical management direction when the program most needed them, and assuredly ranks in that small pantheon of leaders without whom Apollo would not have succeeded.

Shea was not the only Apollo manager who felt responsibility and remorse for not being more aware of potential fire danger. I felt it myself. My confidence and optimism were weakened by this evidence of tunnel-visioned failure to look beyond my immediate concerns and action items. Were we all so busy that we could no longer think? The fire gave objective evidence of our individual and collective shortcomings that no systems and procedures could hide. Did I, and did the Apollo team, have the wisdom, judgment and skill it would take to reach the Moon safely? After Apollo 1 the question was raised anew in our minds, and the answer was not reassuring.

In my new job in S/CAT I was responsible for building and testing the LMs, and I spent much time on the floor in the LM Assembly clean room, or in the subassembly shops. There I saw the added work and manufacturing difficulties that the flammability "fixes" entailed. Since LM-1 and LM-2 were to be unmanned, they were exempt from the changes, but from LM-3 upward and for LTA-8, the manned thermal vacuum test article, all the fixes were rigorously applied within the cabin. (In the interest of standardization, some of the changes in materials, design, and manufacturing practices were also followed in the unpressurized areas outside of the LM cabin, including the descent stage.) The new fire-retardant potting took longer to cure than the material it replaced, and the Beta cloth, with its lower wear resistance and flaking, had to be handled very carefully by our technicians to avoid damage upon installation. Combing and dressing the wiring, hand tying it with Beta cloth tape, and swathing the circuit breakers and switches in the Beta cloth booties were fussy, time-consuming operations. Every time the portable X-ray machine was brought into the LM to inspect connectors and junction boxes, all other workers had to evacuate the immediate area. A major reduction in the use of velcro inside the cabin caused an increase in more cumbersome grommets or ties. None of this helped our schedule situation.

Nowhere in LM were the effects of the fixes more concentrated than in the cockpit control and display panels. Containing hundreds of instruments, switches, and circuit breakers, the backs of these panels were crowded with thousands of wires. The panels and adjacent cabin structure were modified to allow neat, orderly routing of wire bundles and to provide added space for potting, clamps, ties, and booties. The finished products were marvels of dense but purposeful packaging.

The final test of the fixes was the flammability test article (FTA), designed and prepared by Sal Salina and his team. This full-sized steel boiler-plate shell of a lunar module cabin was outfitted inside with a full complement of flight-type materials, furnishings, and equipment that had been modified to the new postfire standards. Many of the units consigned to this test were used design verification or qualification test articles, which had already served their intended purpose in the program. The FTA was filled with pure oxygen at one atmosphere pressure (14.7 psia), then pumped down to 5 psia. Relief valves would open if the outward pressure differential reached 5 psi (20 psia), due to fire-induced heating and pressure rise inside the cabin. Several spark-propane igniters were positioned at critical locations in the cabin, including behind the cockpit panels and under the ECS module.

The modified LM passed this test readily. The test conductors were unable to get anything to burn; most locations just smoldered and charred, then went out when the igniters were shut off. After this required phase of the test was officially passed, an overstress test was conducted by placing a large pan of gasoline on the cabin floor. The gasoline fueled a raging fire, its blaze fully enveloping the cabin and popping open the relief valves, emitting columns of flame and smoke. When the FTA had cooled down sufficiently to be opened and inspected, it was found that the wire insulation, switch and circuit breaker housings, potting, and other plastic materials had charred and melted, and some thin aluminum panels had warped and melted. The spacesuits and hoses had local damage but were still functional. Although there was widespread fire damage, nothing much had burned except the gasoline. We and NASA were satisfied that the LM cabin was fire safe.

In hindsight the net result of the Apollo 1 accident was an eighteen-month slip in the CSM schedules and four months' slip in LM. This allowed LM, which had been about one year behind the CSM and Saturn, to catch up. At the time this was not clear to me or anyone else at Grumman, and we fought tenaciously to preserve every day of our schedule despite the many problems and obstacles.

Reorganization and Recovery

Major management changes took place in NASA and North American as a result of the fire. George M. Low, formerly deputy director of the Manned Spacecraft Center, accepted a demotion in rank but a promotion in real authority and responsibility to become Shea's replacement as ASPO manager. He soon instituted the powerful Change Control Board, which had to grant approval before any significant change could be made to the Apollo spacecraft. The CCB, whose members included Kraft, Faget, Bill Lee (later replaced by Bolender),[7] Kenneth Kleinknecht, Deke Slayton, Tom Markley, and George Abbey, met weekly to review and approve changes proposed by NASA or the prime contractors. Low made the final decision on each item, in his habitual careful, cautious, well-reasoned fashion, after patiently soliciting and hearing all viewpoints. The CCB met weekly and many meetings lasted all day and well into the night, but with rare exceptions the premeeting agenda was dealt with. The CCB became George Low's primary tool for controlling the spacecraft program at a detailed level.

Gilruth assigned astronaut Frank Borman to be NASA's on-site program recovery team leader at Downey, and Low spent considerable time there himself, helping North American sort out and correct its problems. In helping them solve their problems, Low formed strong sympathetic bonds with North American's executives and managers that continued beyond the Apollo program.

At NASA's insistence, Harrison Storms was removed as president of North American's Space Division. His replacement was William D. Bergen, an experienced aerospace executive recruited from the Martin Company. Bergen brought with him two associates from Martin: Bastian "Buzz" Hello to run the Florida launch support operation for North American and John P. Healey to manage assembly and test of the first manned Block 2 (LM compatible) command module at Downey.[8] The capable veteran Dale D. Myers remained as North American's Apollo program director.

Bill Bergen took command at Downey; he evaluated and replaced key personnel and reviewed and overhauled organizations and procedures. One major action was to set up personalized teams to assemble and test each spacecraft in the factory. John Healey, former Manufacturing manager at Martin, was the spacecraft team manager (STM) for CSM-101 and the role model for spacecraft team leadership. Tall, dynamic, and confident, Healey assumed total responsibility for his spacecraft and had the full support of Bergen and Myers. He knew the aerospace manufacturing business and was uncompromising on quality but hard-driving on schedules. Best of all, he got results. Borman and other astronauts were impressed and believed that spacecraft 101 would set a new standard for quality, far above that of its ill-fated predecessor 012. This management approach was soon adopted by Grumman when Evans, Titterton, and Gavin appointed STMs to direct the assembly and test of each lunar module.

Apollo flights resumed in November 1967 with Apollo 4, the first flight of the gigantic Saturn 5 booster rocket. It launched an unmanned Block 1 Apollo spacecraft, the refurbished CSM 017. The mission successfully demonstrated Saturn 5 performance, except for the Pogo problem, which had been encountered in other launch vehicles and immediately became the focus of engineering attention at Marshall. CSM flight objectives were fully accomplished, including such critical items as demonstration of service propulsion system (SPS) performance and restart capability and verification of CM heat-shield adequacy for lunar return reentry.

At last it appeared that the Apollo program was on the way to recovery.

12

Building What I Designed

George Skurla, Grumman's director at Kennedy Space Center, entered Col. Rocco Petrone's spacious office, his lips pressed tight and eyes narrowed into slits. Skurla knew what was coming. LM-1, the first flight lunar module, had been delivered from Bethpage the day before, 21 June 1967, and had immediately failed propulsion system leak tests during receiving inspection. He would get a blunt "chewing out" in Petrone's army company commander style.

Petrone, tan, muscular, and still rock solid like the West Point football player he was, rose from behind his large walnut desk to full imposing height and expanded his broad chest, his face twisting into a dark scowl. As NASA's director of Launch Operations for the Apollo program, Petrone ruled everything on the cape with an iron hand, especially the contractors who supported his launches. He was skeptical of Grumman, the newcomer to space and to the Apollo program, and intolerant of the fast-talking New Yorkers with their atrocious accents and streetwise, independent ways. And now he saw that they could not deliver a quality spacecraft.

"George," Petrone said, "what kind of two-bit garage are you running up in Bethpage? That LM you sent us yesterday is supposed to fly in space, but I wouldn't even allow it on the launch pad. Its propulsion tanks and plumbing leaked like a sieve—it's a piece of junk, garbage! You should be ashamed. And four months late besides."

Skurla averted his eyes and pressed his lips so tightly together they turned white. A veteran Grumman flight test engineer, he had directed the structural flight test demonstration of many Grumman airplanes, including the F9F Panther and Cougar jet fighters, and the large E2C Hawkeye with its dominant rotating radar antenna. He was intensely proud of Grumman and felt that he personally carried on the tradition of the "ironworks." LM-1 was an embarrassment, and he felt betrayed by the people at home.

"Now take it easy, Rocco," Skurla said. "It has some problems, but it's not junk. Nothing with the Grumman name on it is junk."

"When they turned on those sniffers there were sirens wailing everywhere," Petrone continued. "What kind of so-called tests did they do in New York before sending this wreck to us? You guys were supposed to be a cut above North American, but now it seems you're even worse. I want you to tell Evans and Gavin that NASA won't stand for this. They had better get this fixed, and fast! Your name is mud around here until they do."

Skurla retreated to his office and called Gavin. "Joe," he said, "I just sat through the most uncomfortable meeting of my life. Rocco Petrone chewed me up, down, and sideways, and you know something—he was right. You guys really let me down, every one of you. Especially that guy Kelly—wait 'til I get hold of him. How could he send a spacecraft that leaks like a sieve? 'Junk, garbage,' that's what Petrone called it."

Later Skurla was on the phone to me. I had to hold the receiver away from my ear, he was yelling so loudly: "Listen, Kelly, I want you to know you've disgraced the Grumman company, and made my life miserable down here besides. Petrone told me you sent us 'junk, garbage—leaks like a sieve.' What the hell is the matter with you guys? Don't be too proud to get help—talk to Radcliffe, he knows all about leaks. Get Corky Meyer and Flight Test to help you. Don't keep everything locked up tight within Engineering."

"I'm sorry, George," I said. "I'm really sorry. All I can say is it didn't leak when it left here. We must be doing the leak testing differently—we'll have to straighten that out."

"You'd better straighten everything out, and fast! I want this fixed, and I'll get Gavin and Titterton and Evans and God Himself if necessary to make sure it happens. So smarten up, kid. Get off your ass and do what it takes to fix those leaks. And one other thing—this had better never, never, ever happen again!"

Skurla hung up abruptly, leaving me to wallow in gloom and self-recrimination. My S/CAT team and I had let Grumman and Apollo down. We were a laughingstock, the butt of scorn, pariahs of the program. Yet just a few short days ago we had worked our hearts out on LM-1 and were proud of passing NASA's high hurdles for delivery at the customer acceptance readiness review. What had gone wrong?

Shakeup

It began with a fierce snowstorm that struck Long Island in the darkness on Tuesday morning, 7 February 1967. I could hear the wind howling outside, pummeling our bedroom, which was on the storm-targeted northeast corner of the house. It was snowing hard, with perhaps six inches on the ground already. I dialed the Grumman snow closing number and a smile of relief crossed my face as I heard the recorded voice declare that all Grumman plants were closed due to the storm. Hooray! Instead of battling traffic at 7:00 A.M. I could crawl back into my warm bed and look forward to a day of playing in the snow with Joan and the children. I drifted off to sleep again without even a twinge of conscience.

I was exhausted from an endless succession of six- and sometimes seven-day weeks, with twelve- to fourteen-hour days. The workload was very heavy now that much of our system and component hardware was being tested at the suppliers and encountering failures, and our first flight LM was in manufacturing and test, along with other flight LMs and test articles. But beyond fatigue was the shock and dejection that our LM team felt when, just ten days previously, we had learned of the tragic fatal fire in CM 012 on Pad 34 at KSC. The whole program was still reeling from its impact, and there was a degree of confusion as NASA upper management wrestled with the implications of this accident and the extent of corrective actions and program rescheduling that would be required. I was relieved to gain an unexpected day off.

Back at Grumman the next day I was summoned to a special meeting called by George Titterton and Joe Gavin. All the upper management of the LM program was there, and much of Grumman's top corporate management. Titterton sat alone at the plain plastic-topped metal table in the large Plant 25 conference room in a long-sleeved white shirt and dark tie, staring balefully at the expectant audience through his thick glasses. The crowded room fell silent as he began to speak: "I hope you all enjoyed your day off. While you were home doing nothing, I was busy thinking, and I've decided what must be done to get the LM program back on track." He paused for dramatic effect, seeking eye contact with me and other program leaders.

"This program is in terrible shape—right now it's headed for failure. Clint Towl and Lew Evans have asked me to get personally involved to straighten things out, and that's what I'm going to do. I've never failed at anything in my life, and I won't allow LM to fail either." He sniffed as his indignation heightened.

"I'm very disappointed with the leadership of this program, and that goes all the way to the top. Can't anyone on LM do anything right the first time, without alibis and excuses? You haven't yet drawn up a schedule that held for more than a week. Every day we hear about new failures during tests, quality problems and just plain 'goofs.' What is the matter with all of you? What became of the Grumman spirit and talent you once had?
Well, I'm going to set things right if I have to turn the LM program upside down and send all its current leaders packing." Titterton's body quivered with outrage as he delivered this threatening message.

"And it's not just me," he went on. "NASA is very upset about Grumman's poor performance on LM. Joe Shea told me he thought Grumman was complacent and technically arrogant—acting like ivory tower knuckleheads, I think he meant. Well, things are going to change, as of this moment. George Titterton is now in charge of LM, and George Titterton does not fail!"

He then announced that I was being removed as LM Engineering manager and put in charge of the Manufacturing and Test operations in the Plant 5 LM final assembly hangar, with Howard Wright as my deputy. We would immediately locate to a command post in Plant 5 from which we would have authority to resolve all problems, whether in Engineering, Manufacturing, or Test, quickly and on the spot. It would be a two-shift, six-day-a-week operation with Wright and me working alternate shifts. Senior leaders from all the operating departments were also assigned to the command post and spacecraft team managers were appointed to lead a dedicated Assembly and Test team for each LM. All this was effective immediately, and the sooner we moved over to the command post the better.

After the meeting Wright and I retreated in shock to my office. I had recently recruited Howard to join me on LM at the suggestion and assistance of Grant Hedrick. He was a creative electronic systems engineer who led the design and development of the complex avionics systems in the A6 Intruder and the E2 Hawkeye. Knowledgeable in digital systems and radar technology, he was gifted in assembling and integrating many individual systems elements into the architecture needed to perform major mission functions, such as weapons guidance or target detection and tracking. He devised algorithms and software to make system components operate in computer controlled harmony, and excelled in troubleshooting and debugging systems problems. He often flew on test flights aboard the airplanes to verify system performance and check newly developed fixes to problems. Sophisticated digital systems engineering that yielded practical results was his specialty, and his reputation within Grumman was soaring. I only succeeded in attracting him because Hedrick thought Wright had a capable avionics team backing him up on the airplane programs, and LM had a more urgent need for strong electronic systems engineering leadership.

Howard Wright was tall and heavy-set, with large, round eyes, a high forehead, and thin, blondish hair. Careful and precise, he did not jump to conclusions until sure he had ferreted out all the pertinent facts and explanations. He had a wide network of associates throughout Grumman Engineering whose capabilities he had evaluated, and he freely drew upon them for help on LM regardless of which Grumman program they were working on. He was polite and optimistic and seemed to have no preconceptions about how space compared with aircraft engineering. Both required expertise, talent, attention to detail, and follow up in execution, which Wright provided in good measure. I was delighted to have him join me on LM—and embarrassed that he'd only been there a short while when Titterton had suddenly changed everything.

We went to Joe Gavin's office, where with his usual aplomb he calmed us down enough that we could start to plan how to carry out Titterton's edicts.

First order of business was to find a home for the command post. We began with a walk through the LM assembly hangar. It was a cavernous, environmentally clean room, about two hundred feet long by eighty feet wide, with thirty-five feet from the floor to the hook on the overhead traveling crane that ran the length of the ceiling. The room, the walls, ceiling, and floor of which were all white, was brightly lit by rows of recessed fluorescent lights in the ceiling. One long wall of the room was punctuated by large, fixed windows at first- and second-story heights through which activities in the assembly area could be viewed without entering the controlled environment. Both long walls were dominated by two-story LM assembly fixtures, movable steel structures that served as work scaffolds, providing ready access to all levels of the spacecraft. The assembly fixtures were also painted white, but they were heavily festooned with cables, plumbing, work carts, and support equipment of various sizes, which generally were black, gray, or brown. The LM under construction was barely visible when surrounded by its assembly fixture.

To enter the assembly area we went into an adjacent locker room, where we donned medical-style white gowns, hats, and gloves, then brushed our shoes in a mechanical cleaner and donned nylon booties. Clear plastic safety glasses completed the attire required for all workers on the assembly floor. A huge blower and air filtration system, hidden behind the white walls, maintained a positive pressure in the entire assembly room so that dirt could not infiltrate inside. To enter we passed through a double set of pressure-sealed doors.

There was no office space on the assembly floor, although there were some small desk areas that the quality inspectors used to process their paperwork. We quickly concluded that the nascent command post should not be on the assembly floor.

Outside the clean room, the adjacent areas in Plant 5 housed the automated checkout equipment, which was provided and operated by General Electric for all Apollo prime contractors. There were three ACE control rooms at Grumman, which allowed simultaneous electrical test and checkout of LMs in three of the assembly fixtures. The control rooms were large enough to house the numerous test engineers and quality inspectors needed to conduct rigorous, complex tests of the LMs. They had a row of test consoles on one wall, with cathode ray tube (CRT) screens providing readouts of the instrumentation data from the test vehicle. A wide range of stimuli and programmed sequences could be generated as required by the test plan, by the large mainframe computer which controlled ACE's operations at the test conductor's command. Two of the ACE rooms contained viewing windows overlooking the assembly floor, a feature which seemed desirable when the plans were laid out but was of limited practical benefit because of the difficulty of seeing anything within the cluttered assembly fixtures.

With the adjacent Plant 5 real estate solidly occupied by ACE rooms and other indispensable equipment, our search for a home for the command post

The mated (ascent stage and descent stage joined) lunar module in the final-assembly clean room. (Courtesy Northrop/Grumman Corporation)

turned to a small group of office trailers in the parking lot of the courtyard behind the main building, near the outside entrance to the LM assembly entry locker room. These prefabricated buildings, known as the Nerve Center, had been set up a few months previously. They provided, in several spartan, low-ceilinged rooms, space in which the managers, foremen, and test teams could meet to discuss their plans and problems and forecast their schedules and requirements. It was, as a New York clothing chain once advertised, "no fancy fixtures—plain pipe racks." Small windows, dark imitation wood paneling, and dirt-tracked vinyl tile floors created a drab, shabby environment. As we looked around without enthusiasm, Gavin, Wright, and I knew that this was where the command post must be. Although lacking in creature comforts, it was close to the action on the assembly floor but free of the cleanliness restrictions, and the trailer arrangement could be readily expanded and modified as our needs dictated. With its location settled, we began planning the staffing and activities of the command post.

The top NASA Apollo program management visited Grumman to review our status and problems and give us some indication of the program's direction in the wake of the Apollo 1 fire. NASA was very unhappy with our continued schedule slippage. LM-1 was plagued by development flight instrumentation (DFI) problems with twenty of sixty-three measurements inoperative, and LM-2 had lost eighteen days against the schedule in the last twenty-nine. We were projecting ship dates of 27 March for LTA-8 and 11 April for LM-1, but nobody, including ourselves, had much confidence in these estimates. NASA was not very informative regarding changes in program plans due to the fire; apparently these were still being formulated.

Building a Team

A few days later I moved over to the trailers, bringing my secretary with me and my desk and file cabinets. I shared a dark cubicle with Wright, whose desk was next to mine. For the first week we worked the day shift together, but then we moved to two-shift operation with an hour's overlap. We experimented with different shift patterns, finally settling on two weeks on days and then nights, but always with frequent interruptions to attend important NASA or Grumman meetings during the day.

I met the people assigned to LM Assembly and Test and found them conscientious, overworked, and, in many cases, discouraged. So many things needed fixing we hardly knew where to begin. There were not enough people assigned with the right skills and training to support the scope of operations we were trying to conduct. Identifying skill needs and staffing them with capable people was a high priority. Enlisting help from the Personnel Department, I set into motion a personnel inventory of the skills, experience, education, and training of our people, and with our supervisors prepared an updated estimate of our personnel and skills requirements. We were soon able to compare needs with current staffing and to begin recruiting internally and externally to fill the gaps.

Assembly and Test was staffed by manufacturing technicians from the Manufacturing Department and test teams primarily from the Engineering and Flight Test Departments. Supporting groups from the other Grumman functional departments, such as Quality Control, Support Equipment, and Procurement, were also assigned to the area. I turned to the leadership of these Grumman departments for help in getting the people I needed. Thanks to pressure from above by Evans, Titterton, and Gavin, they were generally responsive. Within weeks a steady stream of capable new recruits was bolstering our ranks.

When I moved to Assembly and Test there were two LMs in the clean room: LM-1, the first flight LM, modified for unmanned operation in Earth orbit, and LM-2, planned for manned operation in Earth orbit. LM-3, also intended for Earth orbital manned flight, was moved into the assembly area a week later. The first two lunar-landing-capable LMs, 4 and 5, would soon enter the assembly flow. Two full-sized LM test articles, specially equipped and partially outfitted in accordance with their missions, were under construction on the factory floor in Plant 5 not far from the clean room. LTA-3 was a structural test article that had a complete lightweight LM structure but dummy masses in place of functional equipment. It was specially instrumented with strain gages and vibration transducers for the full vehicle vibration test, landing drop tests, and static and dynamic load tests that it would endure. LTA-8 was a thermal vacuum test article that would verify the adequacy of the LM environmental control system, which maintained livable cabin and spacesuit conditions for the crew, and of the thermal design of surface coatings and multilayered Mylar insulation that protected LM from the wide temperature extremes of space and the Moon. It would be suspended inside a huge thermal vacuum test chamber at the Manned Spacecraft Center in Houston with astronauts inside while infrared lamps, heaters, and cold walls simulated the thermal aspects of a lunar mission. Loaded with thermocouples, heaters, and unique wiring, it was an electrician's nightmare, in some respects more difficult to check out than a flight LM.

NASA's reaction to the Apollo 1 tragedy was to make a major change in the materials allowed in the crew cabins of the CM and LM to reduce fire hazards. Every substitute material, while improved in fire resistance, seemed to be inferior in some other characteristic. Strict rules were adopted governing the routing and securing of electrical wires and wire bundles, cleanup after water-glycol spills, and accountability for objects taken into the cabin. Above all these details, a zealous focus on quality control and configuration accountability was set throughout the program.

In meetings with NASA management it was decided to exempt unmanned

LM-1 from most of the changes since crew safety was not at stake. LM-2 presented a problem because most of its wiring was already installed—it would have to be ripped out and redone to accommodate the changes. It might make more sense from a schedule standpoint to leapfrog LM-3, the wiring of which was only partially installed, ahead of LM-2, and recycle LM-2 later or reconfigure it as an unmanned backup to LM-1. After numerous studies LM-2 was redesignated as an unmanned backup and was used to unload some of the buildup testing from LM-1.

The schedule pressure from NASA and our own upper management was unrelenting, yet we had not shown any ability to hold schedules in Assembly and Test for more than a few days. The LM spacecraft; dozens of complex GSE end items; critical manufacturing techniques, such as silver brazing and connector potting; and test facilities, procedures, software, and special equipment were all being developed simultaneously and were encountering many problems. The result was confusion, frustration, and progress that was painfully slow.

I moved to strengthen the test teams that were assigned to each vehicle by adding new talent and weeding out poor performers. Corky Meyer, head of the Flight Test Department and a veteran experimental test pilot for Grumman, was very forthcoming with talent from his department to bolster our test teams. With Meyer's cooperation, some of the most experienced Flight Test engineers were soon applying their talents to the closely related problems of ground testing manned spacecraft.

Where the Action Is

I spent as much time as possible on the LM assembly floor getting to know the people and watching the processes and problems that were plaguing them. This vast antiseptic realm was the pulsing heart of the LM program, a heart that suffered from periodic bursts of arrhythmia when troubles shut things down. I liked to climb up and down the four levels of the assembly workstand, looking at the delicate LM barricaded in protective steelwork and hooked up to hundreds of intravenous cables and tubes. It was all so complex, and so many people were involved on two different shifts—how could we be absolutely sure of getting everything right? Test, retest, and adhere to the strictest discipline and documentation—that was our answer, a mantra that I chanted for reassurance whenever I felt things were spinning out of control.

Sometimes if it was not crowded I would go inside the LM cabin. A technician stationed outside the forward hatch emptied my pockets into a plastic tray and had me list anything I was bringing inside, a precaution against leaving objects in the cabin. Crawling on all fours through the hatch, I stood up inside the crew compartment at the flight station, about the size of a modest walk-in closet. There were usually one or two technicians inside, talking to the ACE station over the intercom as they set switch positions and actuated controls in accordance with test sequences detailed in the operational checkout procedure book opened before them. But sometimes I was there alone and could let my imagination fly ahead to the day when that very LM, that very square foot of cabin flooring where I was standing, would descend to the Moon's alien surface in the final test of all our efforts and dreams. How I wished to be a stowaway in that tiny cabin! It was as close as I ever got to the Moon, but in my mind it was vivid and thrilling nonetheless.

When a lunar module was being moved or its stages mated I tried to be there. These were rare opportunities to see the LM naked, unencumbered by the workstand and connections that usually covered it. When they entered the assembly floor the LM stages were mostly bare gray aluminum structure, with tanks, plumbing, and wire bundles clearly visible. After months of work they metamorphosed into a shining metallic chrysalis, wrapped in silver and gold blankets of mylar insulation. I examined the framework of thin support tubes as technicians carefully fitted the delicate assembly over tanks and electronic boxes, adhesively bonding it to the underlying structure. Plastic "standoff" tubes the thickness of drinking straws were then bonded to the framework, and to these the insulation blankets and the micrometeoroid "bumper" skins were secured with thin washers. Each stand-off installation was hand crafted, and the adhesive bond had be cured for two hours under heat lamps, then pull-tested with a hand-held scale to verify adequate bond strength. An inadvertent stumble or the careless swing of a ladder could easily damage this pick-up-sticks creation, but vibration testing showed that it remained intact under simulated rocket firing, merely deflecting to accommodate the induced load.

I watched the mating of the almost completed LM-1 ascent and descent stages as they approached delivery—it was an exciting moment. The day before, in accordance with NASA's thorough requirements, I observed a simple OCP to proof test the overhead crane before we entrusted it with the precious spacecraft. How rigid, by-the-numbers NASA is, I thought, insisting that we do this silly proof test when we know the overhead crane is routinely checked every six months. Fortunately I kept this thought to myself, so when the six-ton dead weight slowly slid back to the floor as the crane was trying to hoist it I did not have to explain to anyone. The crane was repaired and the proof test repeated—twice—and I silently praised NASA for its caution.

The descent stage was moved into a rolling portable support stand, and the crane was attached to a hoisting frame secured to the upper hatch. As we gawked upward, the ascent stage was slowly lifted straight up out of the opened workstand, while a small army of technicians watched for unwanted, still-attached connections or interferences. We cheered as the stage slowly rose from the forest of steel and rubber, and saw it rotate under the slender cable, enjoying its first limited taste of flight freedoms to come. Gingerly it

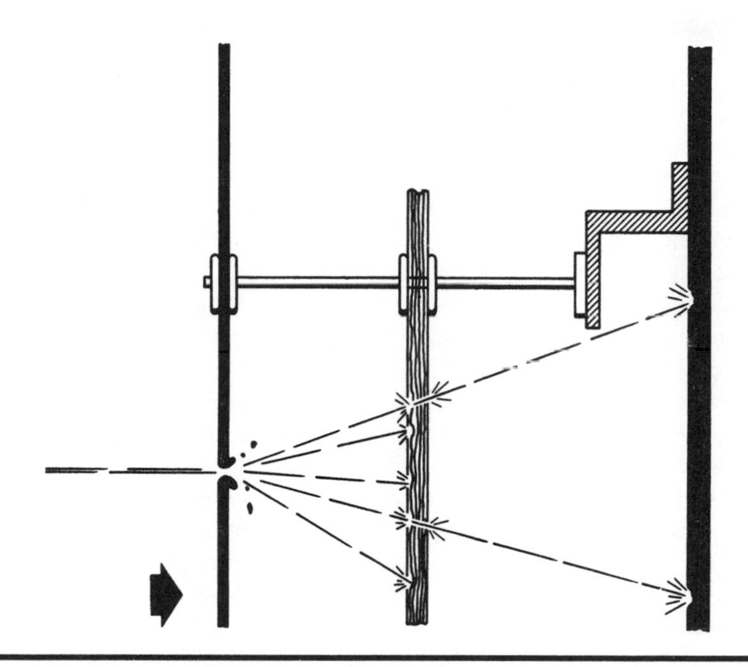

The lunar module's micrometeorite and thermal shields. (Courtesy Northrop/ Grumman Corporation)

was moved over the descent stage and gently lowered into position, eagerly guided to final contact by dozens of upreaching white gloved hands. That night I went home satisfied that we had done something tangible in our journey to the Moon.

Soon after I arrived I encountered a hard-bitten inspector named Dinny Gagnon, who was the chief representative of the Quality Control Department. Dinny, excitable and built like a fireplug, was devoted to producing a quality product and getting things done. One day he told me they were having trouble getting the incoming aluminum alloy sheets for LM through receiving inspection, and he asked me to come with him to see the problem.

We went to a nondescript warehouse-like building in an area of the Bethpage complex that I rarely visited. This was the Receiving Inspection Center. It bordered the Long Island Railroad tracks that bisected Grumman's Bethpage property and was adjacent to the central steam plant from which radiated many large, gleaming silver-insulated steam pipes mounted on short steel suspension poles, providing steam for heat and factory processes to most of the buildings at Grumman Bethpage. The ugly, ubiquitous steam pipes gave the entire complex a look of urban decay combined with technology run amok. The steam plant had a tall industrial smokestack from which thick gray clouds of smoke billowed, and a huge pile of coal outside, astride which a bulldozer was busy smoothing and moving the crumbling black mass.

Inside the Receiving Inspection Center the problem was simply that, with doors constantly open as trucks and trains delivered their shipments, there was no way to keep the area free of the dirt and grit emanating from the steam plant. It was Grumman's microcosm of battery manufacturer Eagle Pitcher's industrial wasteland. The shiny thin aluminum alloy sheets, destined to become lightweight structure for the LM, soon accumulated surface scratches from the gritty environment and were being rejected until the scratches were polished out by hand. Obviously the center had to be relocated to a cleaner building.

I set Dinny loose to find to new home for LM Inspection and swung by Plant 5 for a meeting with Grant Hedrick and some of the corporate engineering leaders. Returning to my office in the trailers later that day I saw Dinny on the phone at his desk, beet red in the face, absolutely livid with rage.

"Whaddaya mean, *your* warehouse?" he shouted in his thick New York accent. "It's a *Grumm's* warehouse, and my boys are comin' in!"

Slamming the phone on the hook he stormed over to me and unleashed an outraged tale of how he had located an ideal area in a clean, lightly used warehouse that was already under the administrative control of the corporate Quality Control Department, but the proprietor, one of his own Inspection peers and colleagues, had the temerity to think he could deny use of the needed space to Dinny. I encouraged him to use my name wherever it might help, and after a few more calls to the Quality Control Department hierarchy Dinny secured a clean new home for LM Receiving Inspection.

Fighting Our Way Through

Parallel, round-the-clock testing of several spacecraft pushed all of us to our limits of endurance and capability. NASA rightly insisted upon rigorous discipline of all assembly and test activities. Nobody could touch a lunar module or its associated GSE without prior approved documentation by work orders, drawings, EOs, test preparation sheets, or test plans, and each step had to be witnessed and validated ("stamped") by a Quality Control inspector. Piles of paperwork were generated, executed, witnessed, and filed on every shift, resulting in the oft-repeated comment, "We don't need rockets to fly to the Moon—we'll just pile up all the paper and walk there."

S/CAT management was vulnerable to sniping because Wright and I, and our spacecraft team managers and supervisors, did not always hear about problems as they occurred on the assembly floor or in the three ACE rooms. In most cases this was as it should be, since the people who could directly resolve the problem always knew about it. But there were instances where problems that required a higher-level decision or approval festered for hours without being resolved. I became convinced that something needed to be done. We emphasized that if anything was holding S/CAT people up that could not be resolved promptly by those involved, it was their duty to work the problem up their chain of command. This worked fairly well for manufacturing, thanks to the encouragement and support of the departmental leadership. The test directors, however, were often too busy with immediate activities and unable to leave their consoles while a test was underway to divert much attention to seeking help. Together we came up with a simple system which alerted management to their problems with minimum effort on the test director's part. We installed a four-position switch at each test director's console, labeled "Off—Running—On Hold—Stopped" and corresponding to green, yellow, and red lights in my and the STMs' offices. Whenever a test was in progress in each ACE room, the lights were lit and we could ask if they needed help to clear a yellow or red condition.

I often answered these silent alarms, usually calling the test director first but sometimes just walking up to the ACE room to see what was happening. Inside the large, dimly lit room, with green lettering flickering on the CRTs at the consoles on one wall, a frustrated group of test engineers and inspectors would be huddled together, looking at diagrams and reading out the steps in the OCP.

A system-by-system buildup of more than seventy operational checkout procedures led up to the integrated all-systems OCP 61018, the most complete and complex of ACE controlled tests. The OCPs began simply, with circuit continuity checks and on/off commands, and progressed to very sophisticated tests that used the powerful ACE computers to simulate the flight mechanics and external interactions of a real mission. One OCP, for example, exercised the rapid-fire sequence of commands that resulted when the LM commander pushed the abort-stage panic button during powered descent. The LM guidance computer (LGC) issued commands that shut down the descent engine, separated the ascent and descent stages, and fired the ascent engine. It made the rendezvous radar find the command module and lock on, while steering the LM toward it. The LGC noted the change in LM's weight and mass properties to ascent stage only, and changed reaction control system commands and trajectory calculations accordingly. On the assembly floor the fluids tanks were empty and explosive device igniters were not installed, so no firings took place, but signals generated by the LM flight hardware were sent to the firing locations in the propulsion, explosive devices and RCS systems and verified. It was an end-to-end electrical simulation of the flight mission event.

S/CAT's ability to hold schedules and do things right the first time was improving, but it was agonizingly slow, always consisting of "two steps forward and one (or more) back." Wright and I benefited from the steadying influence of Joe Gavin and Ralph H. "Doc" Tripp, who helped us analyze what our fundamental problems were and how to cure them, and protected us from the continuous fusillade of "help" from Titterton and others. Doc Tripp, long the manager of Grumman's Instrumentation Department in Flight Test, was brought in early in 1967 as LM program director under Vice President Joe Gavin. He was probably the best manager I ever worked for at Grumman. A firm believer in participative management with tight controls, he began by meeting individually with each person who reported to him and having that subordinate prepare a one page job description and statement of specific goals and objectives. Thereafter he gave each person his head, encouraging his subordinates to take risks and solve problems within their jurisdiction, praising us when we succeeded, defending us when we failed, and welding all of us into a confident LM program team. People- and management-oriented, he made a good counterpart to the more technical and individualistic Gavin. Both of them possessed not only a wealth of good judgment and common sense but also calm temperaments, which allowed them to sail through the recurring crises with equanimity. Tripp had a habit of using his small penknife to whittle designs and figures from wooden coffee stirrers during NASA's long meetings, yet despite repeated attempts, the NASA people never caught him off guard on the topic under discussion. He also refused to adopt NASA's example that the top managers must know every technical system and problem in great detail; Tripp would call upon his experts if pressed on such points.

All too often we in S/CAT would goof or blunder all by ourselves, and no amount of capable LM program leadership could save us from ridicule and upper management or NASA intervention. One of the worst foul-ups occurred on LTA-8, that devilishly complex thermal test article. At one point in the testing we were required to drain and flush the water-glycol coolant that circulated throughout both LM stages, cooling electrical equipment and components and the astronauts' suits and oxygen supply. The test team mounted a fifty-five-gallon drum on the highest level of the workstand, connected by clear plastic tubing to various points of the LM's cooling circuit. Unthinkingly they used an old drum with a rusted bottom that dripped water-glycol solution all over the spacecraft, requiring several days of painstaking swabbing and cleaning with an acid neutralizing agent to set right. At Titterton's meeting I hung my head in apology and endured a bitter, sarcastic tongue-lashing.

The cockpit instruments in the LM had hermetically sealed cases to protect them from humidity and dirt. After we had installed the instruments in three cockpits, a clever NASA inspector at Bethpage devised a new technique that was simple but effective in disclosing dirt particles trapped within the sealed case. He simply held the instrument on the workbench, glass face down, and shook and tapped it gently. Then, retaining the glass-down orientation, he lifted it up over his head and looked for dirt particles on the inside surface of the glass. If any at all were visible, the instrument had to be returned to its supplier to have the case opened, cleaned, and resealed.

This improved inspection technique had never been thought of before, either by the instrument manufacturers or ourselves, and almost none of the instruments that had been delivered could pass it. For the manned flight spacecraft the new test was mandatory, so for LM-3, the cockpit of which was almost complete, we had to remove and replace all the instruments. We faulted ourselves for not devising this obvious quality check long beforehand.

My Chickens Come Home to Roost

My assignment to run S/CAT was a unique learning experience and an appropriate form of retribution. On the assembly floor I came face to face with the troublesome design features I had approved, and in some cases demanded, when I was project engineer, which caused untold hours of toil for the manufacturing technicians who had to make them work in the real world.

Foremost among these were the extremely thin 26-gauge kapton insulated wire and miniature electrical connectors used throughout the LM. Adopted as a weight reduction measure during SWIP in 1966, these fragile wires and tiny connectors were an endless source of problems with wire breakage and improperly mated connectors. Breakage was common in the wires, which were used wherever signal voltages were applied with essentially no current, making them more abundant than the larger-gauge wires.

On the assembly floor I often watched sympathetically as a frustrated technician demonstrated the difficulty of mating and demating a miniature connector containing dozens of fine wires, doing it by feel with gloved hands in a cramped and all-but-inaccessible space. One smiling tech with small but powerful fingers was in great demand because he could handle the most difficult locations.

We ran a special vehicle-level vibration test on LM-1 using electrically driven vibration generators. This test gave us confidence that these wires would not break when the LM was being shaken in flight during launch or from its own rocket engines. Still the wire breakage problem due to installation and handling was a constant drag on assembly and test operations, although it gradually lessened as the technicians improved their techniques for gentle handling of wire bundles. For LM-4 and subsequent vehicles we switched to a special high-strength copper alloy in the wires, which alleviated the breakage problems.

The fire-retarding and moisture-sealing requirements of potting and covering with Beta cloth booties added to the difficulty of handling electrical connectors of all sizes. Much of the potting and bootie installation could be done on the workbenches before the wire harness assemblies were installed into the vehicle, but there were some areas where the wire bundle was threaded through structure and a connector had to be placed on one end inside the spacecraft. The potting material took several hours to cure under heat lamps, after which portable X-ray equipment was used to verify that the wires inside were all properly routed to the connector pins or sockets. A klaxon horn brayed a warning on the floor before the X-ray was turned on—all personnel had to leave the immediate area. Some nights as I lay in bed that klaxon was still reverberating in my brain. If wires had to be replaced in a connector for any reason, they were physically cut out of the potting, the new wires, pins, or sockets were slipped into place, and the connector was repotted and X-rayed again, a cycle that took at least one whole shift. With many thousands of wires and hundreds of connectors in the LM, this was not an uncommon occurrence.

Even the basic aluminum alloy structure of the LM imposed exacting demands upon the manufacturing process. In our relentless quest for weight reduction, we engineers had made widespread use of the high-strength alloy 7075 and of chemical milling, which enhanced susceptibility to stress corrosion. Controlling stress corrosion required carefully fitting each part upon assembly to avoid clamp-up stresses when the fasteners were tightened. This involved educating our mechanical technicians and inspectors to the causes of stress corrosion and the most common fit-up problems of LM parts, and then enforcing rigorous compliance with the approved fit-up procedures. Even so we had numerous cases in which cracks were discovered in thin aluminum flanges that had retained excessive clamp-up stresses when installed. The astronauts were especially vocal in urging us to stamp out this problem, as they no doubt pictured their return vehicle crumbling beneath them on the Moon, victim of its own locked-in stresses.

Delivering LM-1

By early June 1967, despite all these problems, we were approaching the delivery of LM-1 to KSC. As the first flight LM with many special provisions for unmanned flight, the birth pangs of LM-1 were long and painful. The LM mission programmer (LMP), a combination of computer, electrical relays, and switching racks, was the programmable robotic brain that would perform the mission automatically under command of the MIT-designed Apollo computer (primary mode) or the LMP itself (backup mode). This complex, special-purpose unit required extensive qualification testing and design modifications before we began to have confidence in it. Even more troublesome was the development flight instrumentation, a secondary instrumentation system with pressure, temperature, stress, and vibration sensors and its own separate telemetry and recorders. DFI was intended to provide additional engineering data on the early LM flights that could be used to correct and refine the design prior to the first lunar landing; it was only installed on LM-1 to LM-3. Since it was not essential to mission operations, DFI was not qualified to the same high parts selection and reliability standards as the Instrumentation system used for mission critical measurements, but was more akin to experimental aircraft test instrumentation. In mid-February 1967 I noted that 56 out of 320 total DFI measurements were down, with little improvement over recent weeks despite extensive trouble shooting, transducer (sensor) replacement, and wire repair.[1] Many of the DFI measurements were considered mandatory for LM-1's mission of verifying LM systems performance in space.

To cure the DFI problems I got Gene Goltz, head of the Instrumentation Department, to lead a "tiger team" of instrumentation specialists from both the LM program and Flight Test in a complete item-by-item review of the DFI system. They considered component and system design, test results, supplier quality performance, and environmental requirements of the system. Goltz's team recommended strengthening some DFI requirements, particularly the use of vibration testing as a quality acceptance test screening of transducers, modems, and other critical components. They recommended that we drop some suppliers whose components were not performing reliably. Although it took some time, the tiger team's output resulted in acceptable levels of DFI reliability.

Aside from all the problems with LM-1 itself, everything else in the test and checkout operation was simultaneously being developed and tried out for the first time. The ground-support equipment—the cable sets, connectors, adapters, fluid servicing carts, and propellant-loading equipment that connected LM-1 to the computer-controlled automated checkout equipment—also was being designed, built, tested, and evaluated by spacecraft checkout. Hundreds of GSE end items were required, each of which had to be qualified as a deliverable item to the government. Howard Peck, an assistant project engineer, led a team of several hundred engineers and technicians in the development of LM GSE. Even the ACE computers and consoles, developed and maintained by General Electric, were regularly modified and improved as experience was gained in actual test operations.

The test procedures specified each step of test operations, no matter how trivial or routine, and required constant rewriting and improvement to be usable under actual test conditions. It was not unusual at this stage for a test procedure to be rewritten three or four times before a document emerged that could actually be used to conduct the test. With continual interruptions for spacecraft and GSE troubleshooting, repairs, and modifications, test procedure rewrites, and ACE hardware and software upgrades, a nominal six-hour test could take weeks to complete.

That LM-1 advanced through its scheduled sequence of testing at all under such conditions was a tribute to the persistence and drive of the LM-1 test team, and particularly its leader, spacecraft team manager Jim Harrington. Jim was a short, freckle-faced Irishman with a brown cowlick and a bouncy, confident manner. He endeared himself to me during a high-level meeting with NASA's top leadership. George Mueller had asked to meet Harrington, as he was concerned about LM-1's inability to hold schedule. While questioning him in front of the assembled NASA and Grumman top brass, including our president, Lew Evans, Mueller ventured an opinion that LM-1 was having so much trouble getting through its test and checkout phase that it could never be relied upon in flight, and that perhaps it should be converted to a "hangar queen" while LM-2 became the first flight vehicle.

"That's a fine theory, Dr. Mueller," said Jim, staring unblinkingly at him while flashing a beatific smile, "but you're absolutely wrong."

Harrington then outlined the various causes for LM-1 delays, many of which did not even involve the spacecraft itself but would have affected whatever LM was the first to undergo rigorous test and checkout. He ended up convincing Mueller and the other NASA officials that when it was delivered, LM-1 would be a reliable, high-quality product.

Jim's capable leadership and unfailing good humor carried LM-1 forward on the assembly floor too, and by late May it had completed the final and comprehensive all-systems test, the redoubtable OCP 61018. The ascent and descent stages were then demated and moved for pressurized fluid systems testing in the reinforced concrete bunker cold flow facility, located across the parking lot behind the Plant 5 LM assembly hangar, and we scheduled the formal customer acceptance readiness review with NASA for 21 June.

The CARR was such a tragicomic circus that I recounted it at length in chapter 1. Suffice to say here that we did, after a long, difficult day of briefings and "chit" dispositions, receive approval from NASA's Apollo Spacecraft Program Office to ship LM-1 to KSC, subject to cleaning up a long "crab list" of questions, documentation, and minor retests. On 22 June 1967, I stood on the Grumman Bethpage runway in a stiff wind and bright sunshine and watched as the specially designed shipping containers holding LM-1's ascent and descent stages were loaded into the bloated belly of the "Guppy" aircraft, a modified Boeing Stratocruiser used to transport the outsized Apollo spacecraft. With great relief I watched the huge airplane slowly lift off and climb skyward, using most of the six-thousand-foot runway. It was bound for KSC—our first flight spacecraft delivery.

Our relief was short lived. LM-1, scornfully derided as "junk, garbage" by NASA's Petrone, was promptly rejected by receiving inspection at KSC due to plumbing leaks, triggering Skurla's demands for action. A concentrated leak-fix effort led by Will Bischoff, our subsystem engineer for Structures and Mechanical Systems, was successful, but the damage was done. LM-1 had given Grumman a bad name for quality at KSC, a bad name from which we would be slow to recover.

A week after the LM-1 CARR, NASA and Grumman held a CARR on LTA-8. This was much smaller than the LM-1 extravaganza, and it was held in the austere but acoustically adequate conference room in Plant 25. Titterton did not attend, nor did Low; the NASA and Grumman delegations were chaired by Gilruth and Joe Gavin, respectively. This CARR did not result in ap-

Tom Kelly (*left*) and Dick McLaughlin at Grumman-Bethpage for delivery of the LM-1. (Courtesy Northrop/Grumman Corporation)

proval to ship LTA-8 to the Manned Spacecraft Center but generated a long series of action items and discrepancy reports to be worked off. Some of these required lengthy rework in the shop, such as an action to replace all the cockpit instrument panels with another set in which the wiring had the latest fire retardant potting, booties and overcoating. The all-systems test, OCP

61018, had to be repeated after the new panels were installed. The manned LTA-8, in addition to requiring all the changes resulting from the Apollo 1 fire, also had an extensive DFI installation, which had at least as many problems as LM-1's version of DFI.

An Uphill Climb

Gradual improvement in S/CAT's performance took place over the summer, as we worked simultaneously on LTA-8, LM-2 to LM-5, and M-6. The latter was a steel boiler-plate mockup of the LM crew compartment completely outfitted inside with flight-type plumbing, wiring, and equipment, in which ignition tests would be conducted in 5 psia pure oxygen to verify that our cabin was safely fire resistant. In September we completed the action items and DR closeouts for LTA-8 and delivered that unique, nonstandard vehicle to MSC for installation in the large thermal vacuum chamber, where it would perform an extensive, manned test program that was rife with uncommon test conditions and potential safety issues. At least it was out of Bethpage and we could concentrate on building and testing flight LMs!

Another step forward was the official designation of Spacecraft Assembly and Test as a major operational organization within the LM program at Grumman, replacing the temporary and rather vague command-post structure. President Lew Evans kicked off the new S/CAT organization with a rousing speech to the program and the corporate department heads, during which he also formalized the position of the spacecraft directors as leaders of the assembly and test teams for each LM. The spacecraft directors were all highly experienced test engineers, mostly from the Flight Test Department, and one of them, LM-3 director Tom Attridge, was a senior experimental test pilot. Evans challenged the new organization to rapidly improve its ability to increase quality and maintain schedules. I was given carte blanche to add up to fifty test engineers to S/CAT, either by recruiting internally from Engineering and Flight Test or by external hiring. I followed through quickly on this, since I already had a list of internal prospects whom I had intended to pursue as soon as I received budget authority.

The spacecraft directors were a feisty group, accustomed to command positions and pumped up by Lew Evans's charge to straighten out S/CAT, getting it to meet schedules and maintain quality. They worked hard to make their LMs the best of all, and developed a dedicated engineering and manufacturing team devoted to that particular vehicle.

An upturn in our fortunes was the addition to S/CAT management of Lynn Radcliffe, who three years earlier had established a lunar module rocket propulsion test facility at a new NASA site on the White Sands Missile Range in New Mexico. Radcliffe built it into an effective test facility where the LM ascent and descent propulsion systems and reaction control system were test fired under simulated high-altitude conditions, using both boiler-plate and flight-weight test rigs. Joe Gavin wisely observed that Wright and I were wearing down from the unremitting strain of endless deadlines and problems and around-the-clock operations. We needed help at our level, real *help.* The scale of S/CAT operations had grown to involve hundreds of people, all being pushed hard to do quality work but also meet schedules, and personnel and human relations problems were escalating. Radcliffe's greatest strengths lay in human relations, administration and organization, areas that sorely needed strengthening in S/CAT.

After all the internal power plays of the corporate department heads, Radcliffe was a breath of fresh air. He declared that he was working *for me 100 percent,* not as a supernumerary reporting back door to Gavin. With his open features and winning smile, he was outgoing and infectiously enthusiastic—just what we needed. Wright and I spent a full Saturday with him in the trailers getting to know each other—our personality "chemistry" blended perfectly, and we achieved instant trust and rapport. Radcliffe would be deputy S/CAT director of Operations, reporting to me, on an equal level with Wright, who was deputy S/CAT director of Engineering. He took a special assignment to evaluate the strengths and weaknesses of the S/CAT organization and people and develop a plan to "humanize" the operation, which we agreed was simply using its people, not developing or motivating them.

Radcliffe first tackled the problem of assigning key engineers and technicians to individual LM test teams. The test teams were staffed with supervisors and a few key operatives permanently assigned to a particular LM, but the bulk of the personnel required for LM assembly and test was drawn from a pool of skilled people assigned as required for each shift, test, or assembly activity. Since some individuals became recognized as the best in their particular craft, fierce competition existed between the teams to get the "best" people assigned. This competition for scarce resources consumed management time and energy, frequently requiring Wright or me to be the final arbiter.

Radcliffe enlisted Wright to help devise a more efficient approach to personnel allocation. They decided to divide the S/CAT personnel into four permanent teams containing all the skills and manpower required to build and test the LMs. Three of the teams were for the LMs in the S/CAT flow—one in the early phase of assembly and test, one midphase, and one late phase. The fourth team covered test preparations and integration—the operational checkout procedures, test preparation sheets, ground-support equipment validations, and quality documentation without which nothing could be done on the S/CAT assembly floor.

To finalize and implement the "full team" approach, we summoned all the spacecraft directors and team leadership to a marathon "football draft" meeting, at which Radcliffe and Wright explained the concept and presented their version of the "draft picks," with 110 to 120 people named to each team. After a day of haggling, a final version of the list was solidified, with the proviso that informal loans of individuals could be worked out between spacecraft directors at any time if both sides agreed. This new arrangement proved more efficient, largely eliminating the disputes over people assignments and developing a healthy esprit de corps and competition between teams. When a lunar module was delivered to KSC some of the team went with it, while the rest of the people were recycled into the team of the new LM just entering the S/CAT flow.

Radcliffe was also concerned that most S/CAT personnel were widely scattered over the Grumman complex, which made it more difficult to develop a cohesive team spirit. Most S/CAT people rotated periodically onto the assembly floor in Plant 5 to perform scheduled tests or assembly operations. When not required on the floor, they returned to their office or shop areas to hold meetings, review results, prepare for the next activity, or complete documentation. As S/CAT grew into a three-shift, seven-day-a-week operation with fourteen to fifteen hundred people, its people were housed in eight different Grumman Bethpage plants. A larger building to house S/CAT people was urgently needed.

Patiently probing the Grumman corporate bureaucracy, Radcliffe located an available building across the road from Plant 5, formerly owned by the printer who did the LM proposal, and persuaded Tripp and Gavin to have the company buy and refurbish it for S/CAT. Because of the time required for the purchase and extensive renovations, this valuable improvement in S/CAT operations did not take place until I was about to turn S/CAT leadership over to my successor.

After an overwrought test conductor broke down sobbing at his console one night and later revealed his concern about medical symptoms that had recently appeared, we established a program of medical examinations for over 465 S/CAT people, to provide a baseline and identify individuals who should avoid stressful assignments or long hours. We set up a training program in supervisory management for all test conductors and managers, since we had thrust many people into managerial positions without any guidance or training. Although they called it "charm school" and downplayed its importance, it was well attended (250 people), and our people took it as evidence that at last somebody cared about *them*.

The Phoenix Rises

Gradually S/CAT's performance improved enough to be noticeable even by NASA. Although a number of technical and manufacturing process problems persisted, they were being identified and solved more quickly and systematically than previously. Test operations were stabilizing as a combination of factors began to work in the positive direction. After many rewrites the test preparation sheets and operational checkout procedures were usable and stable. The test teams had gained experience and confidence and were augmented in numbers and skills. The GSE was more operationally ready as design errors and omissions were corrected and installation problems resolved. The net result was that our ability to hold schedules was improving steadily: major OCPs could now be completed in days rather than weeks.

The unmanned LM-1 was launched into Earth orbit in late January 1968 as the Apollo 5 mission, and it achieved a qualified success. It met all its major mission objectives but had to do them in a backup mode, flying under control of the LMP. Primary mode, the LM guidance computer, could not be used to control the propulsion burns because of a software error that caused premature engine shutdown. Everything else on LM-1 worked as planned, so it appeared that a second unmanned flight using LM-2 might not be necessary. The Apollo 5 flight records had to be carefully studied before this decision could be made.

While I was defining the analyses needed to decide the plans for LM-2, I found out that Joe Gavin was entertaining the possibility of pulling me, Wright and Radcliffe out of S/CAT and into the LM program office. Gavin felt that we had succeeded in turning the S/CAT operation around and in laying a firm foundation that others could build upon. We were needed to strengthen the program office in a number of areas. The technical problems that dogged the LM, and new ones which constantly turned up, had to be resolved quickly through George Low's formal NASA Change Control Board process. Low used a weekly CCB meeting at Houston, Downey, or Bethpage to debate, approve and settle the corrective actions resulting from test failures and design and manufacturing process deficiencies. This was a management-intensive activity, since every case had to be carefully prepared and briefed to NASA's upper management, which made up the CCB.

Plans for the missions following the first lunar landing required increased attention. It looked possible for LM-5 or -6 (Apollo 11 or 12) to be the first landing, but hardware and crews were in the pipeline for missions up to Apollo 18.[2] What would be done on all these subsequent missions? The Wernher von Braun team promised that Saturn 5 performance would be improved somewhat on the last three or four missions—how should this extra launch payload be used? Modifying the LM to increase its lunar surface stay time and scientific equipment payload was a possibility for the later missions.

At a NASA Apollo program meeting in headquarters in early February 1968, the decision was made that an additional unmanned LM flight was not necessary. The next LM flight would be manned, using LM-3 in an Earth orbital rendezvous mission with the command-and-service modules. LM-2, configured like LM-1 for unmanned flight, would be completed, formally delivered to NASA via a DD-250 (the form by which the government accepts delivery and ownership of a product from a contractor) and loaded into its shipping containers for storage. LM-3 would receive top priority in S/CAT plans and schedules.

LM-2 was delivered to NASA on 17 February 1968, only six days behind the contract delivery date. S/CAT had finally shown that it could set schedules and hold them. Two days later I announced to my S/CAT team that I was being reassigned to the LM program office as assistant program director–Engineering, and that Howard Wright and Lynn Radcliffe were also being reassigned to LM program positions. Paul Butler, a veteran Flight Test engineer and manager, succeeded me as S/CAT director. It was a bittersweet moment to leave the people in manufacturing and test operations who had come to mean so much to me and whose commitment to the success of the program was unstinting. As a culmination of our revitalization efforts, within a few weeks S/CAT headquarters would be moving into the newly refurbished building across the street from Plant 5.[3]

Thus ended my most challenging, frustrating and rewarding year on the Apollo program. S/CAT brought me totally down to reality—down and dirty with the thousands of physical details that had to be perfectly crafted, installed, verified, and documented, and face to face with the earnest, hardworking men and women who strove to do their very best to build a spacecraft that would land men on the Moon and bring them back safely. Thousands of devilish details and the limits of human fatigue and frailty stood in our way. I had seen the effort and concentration by hundreds of skilled craftsmen that was needed to make Engineering orders or program decisions take shape in fact, not just on paper. I had successfully transformed a group of talented but fractious individuals into a smoothly functioning team, capable of performing with harmony and professionalism like a symphony orchestra. I could now move on to even greater challenges, as the end of the decade that the late president had set as the nation's lunar landing goal was fast approaching and we were not yet on the Moon.

3

Flying

13

First LM in Space

Apollo 5

In January 1968 the Apollo 5 mission with the unmanned LM-1 aboard was almost ready to be launched into Earth orbit from KSC atop a Saturn 1B booster. I got a welcome break from S/CAT to support the mission at Kennedy Space Center and took Joan and our three oldest children with me to watch the launch. We arrived a day ahead and checked into the Howard Johnson Motel on the oceanfront at Cocoa Beach.

After negotiating the daunting security check-in at the KSC gate, I drove onto the vast semitropical sandspit where NASA's facilities were located and found the sprawling, three-story Operations and Checkout (O&C) building. At the front of the building, which faced the road, there were three long, horizontal stripes of windows where the office areas were located; at the rear were the white-sided high-bay areas that housed the spacecraft assembly and test clean rooms, altitude chambers, automatic checkout equipment stations, and simulators. Inside the Grumman office area on the second floor I was greeted by Herb Grossman, our KSC Engineering manager, who showed me around, introducing people if I did not already know them and filling me in on the status of the mission. Herb was a study in perpetual motion, a dynamic man with the confidence of a natural leader. He was of medium height and solidly built, with a chiseled face, dark wavy hair, and a firm, authoritative bearing. He had a broad systems engineering background in analysis and test and was also very attuned to the nuances of internal office politics. Grossman often explained to me subtle rivalries between individuals and organizations to which I had been oblivious.

We toured the Grumman office area, and later Shaler "Hobie" Gilman took me on a tour of the Vehicle Assembly Building (VAB), the enormous building in which the complete booster-spacecraft "stack" was assembled atop the massive launch transporter, which would move it out to the launch pad using its four sets of giant tank-tread crawlers. Gilman was a seasoned

KSC veteran, having worked in launch operations for Convair before joining Grumman, and was one of the reasons the Grumman team learned the ropes at KSC so quickly.

The VAB is one of the wonders of the modern world. Said to be the largest manmade structure on Earth, it looms for miles like a massive black and white temple of technology over the flat eastern Florida coastline. Gilman took me up in an elevator to the 330-foot level, where the command module would be positioned atop the three-stage Saturn 5 booster and the spacecraft/LM adapter. There was no launch vehicle in the cavernous bay, and we peered over a guardrail at the antlike people walking on the ground floor far below. We then went out to Launch Complex 39, which was used for Saturn 5 launches, and was the spot from which Apollo would someday leave for the Moon. From a lower level of the thirty-three-story launch tower we looked at Apollo 5 on Launch Complex 37 about three miles away. Shimmering in the distance and surrounded by its launch tower, the two-stage Saturn 1B had a massive, stubby look, accentuated by the lack of an Apollo CSM spacecraft and launch escape tower atop the spacecraft/LM adapter. As in all Apollo launches, the LM was not visible, being nestled inside the structural shell of the SLA.

The Grummanites whom I met on this brief tour were generally upbeat and confident, but also somewhat apprehensive. "Keep your fingers crossed" was a frequent admonition. I understood their concern, as this was the first time the Grumman launch support team had been put to the test in a launch operation, with all its staggering complexity and irreversibility. NASA and the other aerospace contractors at KSC were watching closely to see how the newcomers from Long Island measured up to the unforgiving discipline of an Apollo launch.

Back at the O&C I ran into George Skurla, Grumman's KSC director, making his rounds on the Engineering floor. George greeted me warmly; he welcomed all the help he could get, especially for this crucial baptism of fire. We had succeeded in stopping the leaks in LM-1, and he held no residual grudge against me or my organization: there were too many more launches ahead to waste time looking backward. Skurla had previously headed Grumman's Structural Flight Test Group and was initially hesitant to accept his new assignment in Florida, but now he was reveling in the leadership and management challenge of his job.

Skurla was tall and trim with wavy dark hair. His eyes were close-set and darting, as though in search of hidden enemies, and he wore a morose, worried expression, except when he smiled, which was not infrequently. He was a born worrier and assumed a conspiratorial air with his listeners when expressing his concerns, huddling with them and lowering his voice for emphasis. His worries drove him to anticipate problems and try to prevent their occurrence, an excellent habit for anyone directing rocket launch operations.

Skurla was also a stickler for detail and rigor, a natural bent reinforced by his experience with the unforgiving nature of aircraft flight testing. These characteristics enabled him to earn the grudging respect of his NASA counterpart and overseer, Col. Rocco Petrone, who ran a disciplined, military-style operation at KSC.

Because of the short duration of the Apollo 5 mission (less than eight hours), I planned to support the mission from KSC, even though it would be directed from the Mission Control Center (MCC) at NASA-Houston once the booster rocket cleared the launch tower.[1] I was in an unofficial Grumman LM spacecraft support room in the O&C, which had mission monitor displays and the Apollo audio network. Several other Bethpage engineers were there with me to support the mission, including Bob Carbee and Manning Dandridge, but most of our mission support engineers were in Houston, led by John Coursen, Engineering manager, and his deputy, Erick Stern. We at KSC would funnel our comments and problem responses into Coursen and Stern.

The next morning the delicious tingle of anticipation hung over KSC like the morning sea haze through which the fiery orange Sun groped its way to the ground. The roads were clogged with thousands of spectators, many of whom had camped on the beaches overnight. Coastal road A1A became a vast parking lot as crowds of partying visitors in baseball caps, flowered shirts, shorts, and dark glasses jostled for space. The launch was scheduled for noon, we were told to report to the VIP viewing area at 10:00 A.M. About two miles away across a flooded tidal lagoon, Apollo 5 brooded on the launch pad.

Joan and our three boys were thrilled to be there and dazzled to see numerous celebrities, including vice president Hubert Humphrey. The launch countdown did not go smoothly—there were frequent unplanned "holds" while the launch team coped with problems, of which LM-1 was the focus. The water boiler temperature in the LM coolant loop rose out of limits, due to a problem in the ground-support equipment (GSE) freon supply, and a power supply in the ground-based digital data acquisition system failed and had to be replaced. Hours ticked by, and the crowd became bored and restive. Our kids played games among themselves and explored the edges of the tidal lagoon until a security guard chased them back into the grandstand area. The sun set in a flaming red orange ball while purple shadows gathered over the ocean as the launch countdown neared its end.

It was the first space launch we had seen firsthand, and it did not disappoint in spectacle and beauty. First we beheld the brilliant orange flame of the Saturn 1B, then the agonizing wait until the hold-down clamps were released and the rocket began its slow climb upward alongside the launch tower, finally clearing it. Then came the heavy, deep-throated roar of the mighty engines, simultaneously pressing down from the sky and upward, like an earthquake, from below the ground. Set majestically against the rose, purple, and deep blue of the dusky sky, the blazing torch of the rocket lit up the approaching night for miles around. It was a thrilling sight but also reminded me of the inherent risk of our whole enterprise. So much raw power, unleashed in such a short time! The awed crowd dispersed, and I dropped Joan and the boys off at the motel and drove to my post at the O&C building.

Apollo 5 had an ambitious set of mission objectives. In addition to verifying the satisfactory operation of LM's systems in space and its controllability and maneuverability as a flying machine, the mission concentrated on demonstrating the performance of the ascent and descent propulsion systems, and its ability to perform an abort-stage maneuver. This time-critical event occurred if it were necessary to abruptly abort the powered lunar descent. It required simultaneously shutting down the descent engine, separating the ascent and descent stages, and igniting the ascent engine. The ascent engine would start while still atop the descent stage, as in a liftoff from the Moon, and its exhaust would initially impinge upon and be deflected by the top surface of the descent stage, a condition known as "fire in the hole," or FITH. There was some concern that in an abort-stage maneuver the aerodynamic forces of FITH might cause the descent stage to tumble, since when separated from the ascent stage it had no attitude control. A tumbling descent stage could possibly impact the departing ascent stage. The only way to put this concern to rest was by Apollo 5's flight test demonstration.

LM-1 was put into a perfect orbit by its Saturn 1B booster, Apollo-Saturn 204, the same rocket that had been assigned to launch the ill-fated Apollo 1. The spacecraft/LM adapter and the nose cone were jettisoned as planned, and the unmanned LM's systems were activated by ground commands from Mission Control in Houston. After checking out the spacecraft for two orbital revolutions, Mission Control commanded the first major systems demonstration of the flight; a thirty-eight-second firing of the descent engine. However, after four seconds LM-1's guidance system shut down the engine because it had not provided sufficient acceleration to satisfy a cutoff threshold built into the software. Consulting with my guidance experts at KSC, Houston, and Bethpage, we quickly established that this was simply a software error that did not indicate any problem with the descent propulsion system. In a normal descent propulsion start-up the propellant tanks would be fully pressurized before the firing command was initiated, and the acceleration would exceed the required threshold in less than four seconds. However, LM-1's first descent propulsion firing began with the tanks only partially pressurized as they were at Earth launch, so it took an additional two seconds to reach full tank pressure, engine thrust and spacecraft acceleration. The software cutoff should have been changed to six seconds for the first firing, but it was not, causing the unwarranted engine shut down.

NASA Houston agreed with our conclusion that there was nothing wrong with the LM, and switched to a backup plan using ground commands to the LM mission programmer to salvage the mission. Continuing with the preset in-flight commands that had been programmed into the LM guidance computer's software was no longer feasible because the timing of the mission events had been upset by the need to repeat the aborted descent engine firing. Event timing was critical in Earth orbit because the spacecraft must be in contact with a ground station of the worldwide tracking network whenever flight activities occurred. The LMP, a special piece of equipment for unmanned LMs only, allowed LM systems operations and maneuvers to be commanded from the ground in any desired timing and sequence.

Transitioning smoothly into this well-rehearsed fallback mode of operation, Flight Director Gene Kranz's firm, confident voice over the net led the rearrangement of mission events to assure that, if the LM were capable, all flight objectives would be met. Kranz and his flight controllers consulted frequently with the Grumman support staff, and we all concurred on their revised flight plan. Houston executed the plan crisply and successfully performed all mission events, including prolonged descent and ascent engine burns, and FITH ascent engine start-up and separation from the descent stage. After cross-checking that all mission objectives had been accomplished, LM-1's ascent stage was shut down for reentry less than eight hours after liftoff from KSC. It was a brilliant recovery from an uncertain beginning.[2]

I was elated by LM's basically good performance in its first flight in space. Witnessing a launch brought home to me the full complexity of Apollo missions, and how totally dependant they were on every one of the thousands of people involved doing his or her job right. Trust in the quality of the work performed extended far upstream of the final assembly process and down to the final launch preparations at KSC. Apollo's successes would be the proud result of the individual efforts of thousands of engineers, managers, crafts people, and artisans, but a critical moment of carelessness or inattention by any one person could cause failure. It was a sobering thought as we prepared to move into manned LM flights.

14

The Dress Rehearsals

Apollos 9 and 10

The fateful year arrived—1969, the final year of the decade chosen by President Kennedy for America to make history in space and show its technological supremacy over the Soviet Union. After missing Apollo 8 the LM still had not been flown by astronauts in space, but NASA plunged ahead bullishly, counting on success, and scheduled launches every two months beginning in March. Three flight LMs (LM-3, LM-4, and LM-5) were at KSC undergoing checkout and prelaunch preparations. The high-bay spacecraft assembly clean room in the O&C building pulsed with around-the-clock activity: three command modules and service modules were on the floor also, tended by their own legions of engineers and technicians from North American.

First Manned LM Flight: Apollo 9

The tardy LM was the program's big unknown. Data from the unmanned flight looked good, but LM lacked the assurance that comes from having sharp-eyed astronauts living aboard in space, flying, probing, noticing every detail of its in-flight performance, up close and personal. Wally Schirra and Frank Borman and their crews made many observations and suggestions in their debriefings from the Apollo 7 and 8 missions that helped NASA and North American make small improvements to the spacecraft, equipment stowage, and mission operations techniques that bolstered confidence in the CSM's readiness to head for the Moon. LM needed this also, and it had to show that it could perform the critical mission functions under pilot commands.

Building and Testing

LM-3 was designated to be the first manned LM, and was first to receive all the materials changeouts, quality procedure enhancements, and design changes resulting from the Apollo 1 fire. This slowed its progress in Spacecraft Assembly and Test in Bethpage, but by fall of 1967 the basic manufacturing and assembly was completed and the ascent and descent stages were mated together in the combined stages workstand. Further equipment and component installations continued, and the spacecraft was connected, by white "interface boxes" on the assembly floor and workstand platform levels and many thick, sinuous black cable bundles, to the automatic checkout equipment station that overlooked the floor of the clean room. This allowed formal acceptance testing of LM-3 to begin even as further manufacturing assembly continued.

As LM-3 moved further into the checkout sequence, the tests required interaction with the pilots in the LM cockpit. Grumman test engineers and LM project pilots Jack Stephenson and Scott MacLeod fulfilled this function, which could be required at any time during S/CAT's three-shift, twenty-four-hour workday. NASA's flight crews intended to participate in these tests, and we set up trailers in the parking lot behind Plant 5, adjacent to the high-bay spacecraft Assembly and Test clean room, as their office and motel. For each manned Apollo flight, NASA assigned three crews: the prime crew, who would fly the mission unless ill or incapacitated before launch; the backup crew, who would assist the prime crew in preparing for the mission, representing them at meetings, briefings, inspections and tests, and substituting for a prime crew member on the flight if required; and the support crew, who represented both the prime and backup crews as needed to cover the many simultaneous program events and activities in which the crews were interested. For Apollo 9, the crew assignments were as follows:

Crew	**Commander**	**LM Pilot**	**CM Pilot**
Prime	James McDivitt	Russell Schweickart	David Scott
Backup	Charles P. Conrad	Alan Bean	Richard Gordon
Support	Edgar Mitchell	Jack Lousma	Alfred Worden

We saw mainly McDivitt, Schweickart, Conrad, and Bean on LM-3. The flight crews spent more and more time at Bethpage, sometimes standing for hours in the LM crew compartment as an OCP slogged through its lines. They delved into their LM with intensity: questioning everything, reviewing equipment test records from the suppliers, visiting key subcontractors, and acting as the scripted pilots in major tests. Their curiosity, persistence, and endurance knew no bounds.[1] Stephenson and MacLeod helped them at Bethpage, performing tests for them in the LM cockpit and following up to obtain answers to their lists of questions.

Jim McDivitt was a good-looking fellow, above average in height and with a trim build. He had a pleasant face, a ready smile, and quizzical eyebrows that could point upward toward the center of his forehead in silent incredulity when offered an unsatisfactory explanation. He was a veteran astronaut, having been commander of Gemini 4, a trailblazing flight that included a rendezvous experiment and a highly successful rendezvous exercise.[2] He had also been the backup commander for the ill- fated Apollo 1. Jim had a wide understanding of aerospace systems and design and argued tenaciously with me and my engineers about any aspect of LM that did not seem right. He roamed the assembly clean room and shop floors widely when at Bethpage, poking into obscure corners and questioning the workers about what they were doing and why. Although unfailingly polite, Jim would not tolerate excuses or evasions.

"Rusty" Schweickart was a rangy, rawboned young man with short, reddish hair, a ruddy, freckled complexion, and the open, earnest expression of one eager to learn and perform. Rusty wanted to know absolutely everything about his LM, and he patiently studied system diagrams, operating manuals and test procedures until he understood how each system and component worked. Then he could recognize normal system performance and visualize operations in backup and degraded failure modes. As the LM pilot he spent many hours in the LM mission simulators at Houston and at Kennedy Space Center, learning to fly the LM in all mission phases and failure conditions. McDivitt had to fly both the lunar module and the command module simulators, but Schweickart could concentrate on the LM, becoming an expert pilot, able to handle any emergency.

Simulators and Combat Boots

The mission simulators for the LM and the CM were workhorses of the astronaut training program. Both were made by Singer-Link, successor to Link Aviation, which made the famous Link Trainer, the rudimentary cockpit flight simulator that was the first step toward the skies for thousands of World War II pilot trainees. We placed the LM simulator under contract a year and a half after Grumman's go-ahead, once the preliminary design of the spacecraft firmed up. It was a complex computer-controlled flight simulator with accurate replication of the LM flight station, controls, and displays. Realistic lunar surface scenes were projected onto the windows by an optical system developed by Farrand Optical Company. This innovative system used a small fiberoptic camera that "flew" over a three-dimensional plaster model of the lunar landing site (or the Earth for Apollo 9), remotely driven by computer commands that matched the LM's flight path.

The LM mission simulator could insert faults into the LM's systems, displaying the resultant instrument readouts, caution and warning alarms, and effects on LM's operation and flight performance. Bogus failures were entered into the LMS from an instructor's console outside the simulated crew compartment. It was a fixed-base simulator—there was no motion of the crew compartment, but an electrical vibrator mimicked rocket firing and audio tapes played cabin noises, such as environmental control system fan and pump hums and reaction control system thruster firings. The LMS could replicate LM's communication systems accurately, providing blackouts from contact with Earth when LM was behind the Moon and static and dropouts if LM's steerable antenna lost lock or the omni antennas were in unfavorable positions. There were two LM mission simulators, at MSC Houston and KSC Florida, and they were booked weeks ahead for flight crew training and mission simulation support.

Grumman also designed and developed the full mission engineering simulator (FMES), an in-house Grumman guidance and control system simulator used for engineering development and integration of the LM's guidance, flight controls, and computers. The FMES had a rudimentary flight station that did not stress realism and lacked optical displays, but it had a highly accurate computer-controlled flight attitude table on which inertial measurement components (platforms, gyros, accelerometers) were mounted and subjected to flightlike environments. Prototype flight hardware was used in these tests, and during flight missions the FMES proved a vital problem-solving resource.

One Sunday winter morning McDivitt came charging into my small office in a trailer behind the LM Assembly Hangar in Bethpage and shouted, "Hey, Kelly! Do you know there are guys wearing combat boots in the LM cabin? They could put their feet right through the cabin skin if they're not careful."

I followed him into the LM assembly hangar. Everyone entering had to don white smocks, caps, gloves, and booties, but we allowed them to put the cloth booties on over their shoes (not expecting boots!). McDivitt was right; I found at least three boot wearers myself. We changed our factory dress rules to require that shoes be left in the entry area and the booties put on over socks.

Passing Muster at KSC

After passing an intense two-phase customer acceptance readiness review at Bethpage, LM-3 was shipped to Kennedy Space Center on 14 June 1968—my birthday, which I took to be a good omen. To prevent a repeat of the LM-1 receiving inspection debacle at KSC, we arranged to have a team of NASA and Grumman quality inspectors work at Bethpage for several weeks before delivery, sharing inspection duties with the resident team in S/CAT. Both inspection teams signed off on the delivery acceptance papers. (LM-2, not required for flight because of LM-1's success, was delivered to NASA for storage and ultimately displayed in the Smithsonian National Air and Space Museum in Washington, D.C.)

Despite this precaution more than one hundred discrepancies were found in LM-3 during receiving inspection at KSC. Broken wires and structural cracks due to stress corrosion were of the greatest concern. LM-3 contained thin, 26-gauge wires with low-strength annealed copper alloy, which were very delicate. The failure of a window on LM-5 during a pressure test at Bethpage cast further doubt over LM-3, prompting intense inspection of its windows while the engineering investigation proceeded at Corning Glass.

NASA's concern was so great that after a month of inspection and tests at KSC, they had George C. White, chief of Reliability and Quality Assurance at NASA Headquarters, personally inspect LM-3 and review all the findings. He identified nineteen areas in which the craft had quality problems requiring evaluation by the Certification Review Board before clearance for flight. The resulting uncertainty over when LM-3 would be ready for flight caused NASA to consider alternative missions to prevent too long a gap between launches. The first manned orbital flight of command and service modules, Apollo 7, was scheduled for October 1968, to be followed by a manned CSM/LM flight demonstrating rendezvous in Earth orbit in December. George Low proposed delaying the latter mission because LM-3 would not be ready and substituting a CSM-only lunar-orbit mission. He sold this idea to the NASA hierarchy, and it was officially adopted.[3]

Air force Brigadier General Carroll H. "Rip" Bolender, LM program manager at NASA-Houston, announced the decision to us in Bethpage. I was embarrassed that NASA had to improvise a mission because our LM was not ready, but I was thrilled with the mission they chose. Orbiting the Moon at Christmastime with Apollo 8 seemed an inspired choice, and it would settle some nagging concerns about the accuracy of lunar-orbit navigation in the complex, "lumpy" gravitational field of the Moon.[4] General Bolender told us NASA wanted Grumman people to observe the mission support operations on the new mission, designated Apollo 8, as preliminary training for our first manned mission. When the time came, John Coursen, Bob Carbee, Arnold Whitaker, and other key engineers observed mission support operations on Apollo 8 in Houston. I stayed in Bethpage to coordinate solutions to problems with predelivery operations on LM-5 and LM-6 and to make prelaunch preparations for LM-3 and LM-4 at KSC.

LM-3 continued to have problems, especially wire breakage. George Low dispatched Martin L. Raines, Reliability and Quality Assurance chief at Houston, to KSC in January 1969 to assess how bad its wiring was. He found hundreds of wire splices and repairs but considered them safe, and the spacecraft was fully functional and continued to pass its operational checkout procedures. In the other major problem area, stress corrosion, Grumman inspected more than fourteen hundred components on LM-3 to LM-8 and replaced any with cracks. Some of the structural tubes of 7075-T6 aluminum alloy were replaced with the more corrosion resistant 7075-T73 temper. At the LM-3 design certification review at NASA Headquarters in early January, all previously identified issues were declared resolved. The prime crew made their own investigation of LM-3's status, and at the flight readiness review at KSC in mid-

February they concurred that her quality was satisfactory, and she was ready for flight.[5]

I traveled to Houston to take part in mission simulations of Apollo 9. These were more complex and realistic than for the unmanned Apollo 5, since in addition to the flight controllers in the Mission Operations Control Room (MOCR), the stations of the worldwide Manned Spaceflight Network (MSFN), the contractors in the Spacecraft Analysis (SPAN) Room and the Mission Evaluation Room, and other ground-support personnel, they included the astronauts flying the command module and lunar module mission simulators. From my station in the windowless SPAN Room, across the hall from the darkened, theaterlike MOCR with its rows of flight controllers intently gazing at the greenish displays on their cathode ray tube (CRT) consoles, it was indistinguishable from the real thing. I saw the same LM instrumentation readings on the CRT and heard the same network protocols over my headset as in an actual mission, including communication acquisition and loss of signal as the orbiting spacecraft entered and left the range of each ground tracking station, and the capsule communicator (CapCom) conversing with the flight crew. The CapCom, himself an astronaut, was the only person on the net allowed to directly communicate with the crew; all messages from the ground were relayed by him.

Once during a simulation of Apollo 9 this carefully contrived virtual reality was shattered abruptly. During a quiet period in the mission, the LM (mission call-name Spider) failed to reacquire communications with the tracking station at Woomera, Australia, after passing over the Indian Ocean and the Australian outback. The ground team began analyzing the LM's communications systems and could see no problem. Since Spider was flying formation at a distance with CM Gumdrop, CapCom asked Dave Scott, the CM pilot, to hail Spider using Gumdrop's radio. Scott hesitated and delayed for several minutes in following through on the request. Then Comdr. Jim McDivitt's voice came sheepishly over the net: "I'm sorry fellows; you caught us. Rusty and I sneaked out for a few minutes to get a sandwich; we thought we'd make it back before acquiring the next ground station. Sorry about that!"

The simulations allowed us to practice our assigned roles and sharpen our real-time problem-solving skills. As the senior Grumman representative, my job in SPAN was to evaluate each LM problem as it arose and to marshall as much expert help in the solution as time allowed. I collected their inputs, made my own evaluation, and presented this to the NASA senior person in SPAN (usually Owen Maynard or Scott Simpkinson) as Grumman's official recommendations. Any deviation from normal performance was called an anomaly, and each anomaly was written up on a discrepancy report form as it occurred and dispositioned when the explanation of the cause was agreed upon. This might occur weeks later, after extensive computer analyses or laboratory tests had been performed.

The last mission simulation I took part in was about two weeks before scheduled launch. By then we were part of a finely honed mission support team, ready to assist the NASA flight director and his flight controllers in dealing with the unexpected as it came up in flight. We knew the astronauts' lives would be depending upon LM for the first time in space.

Apollo 9 was also LM's first flight together with the command and service modules. It was an ambitious ten-day mission with the goal of performing in Earth orbit the entire sequence of events required on a lunar mission, except for the actual landing. It also provided the first and only flight test of the spacesuit and backpack to be used in exploring the Moon, during extravehicular activity, or spacewalks, from both the LM and the CM.

You Have Twenty Minutes . . .

Before it was even launched Apollo 9 provided one of the most difficult tests of my ability to assess conflicting data and provide NASA with a sound recommendation. About four hours before scheduled liftoff at KSC, an anomaly was noticed during prelaunch filling of the LM's descent helium tank. The helium was used to pressurize the descent propellant tanks, causing the fuel and oxidizer to flow into the rocket engine without pumps. To minimize tank weight, the helium was stored at high pressure and extremely low temperature (nominally 1,540 pounds per square inch at minus four hundred degrees Fahrenheit), which put it in what thermodynamicists call the supercritical state, where it is as dense as a liquid and yet completely fills its container like a gas. This required a tank of advanced design with a highly effective vacuum jacket and insulation, made for us by the Airesearch Division of Garrett Corporation. When filling the tank the servicing crew weighed the amount of helium delivered and verified the tank's thermal performance by reading the predicted combination of temperature and pressure from a thermodynamic chart.

When Spider's helium tank was loaded, the resulting temperature and pressure fell above the predicted curve, suggesting that excess heat was leaking into the helium. Since this could be caused by an insulation defect in the helium servicing cart or its vacuum-jacketed hoses as well as the flight tank, our first recommendation was to empty the flight tank and refill it, using a different servicing cart and hose set. I was at my post in the SPAN Room at the Mission Control Center in Houston discussing the problem with the NASA propulsion people there, as well as on the phone with our Grumman people at KSC. It took more than an hour to detank and refill, and although the result was somewhat improved over the previous, it was not quite within limits. We decided to wait half an hour and see what happened, reasoning that if there were an insulation defect in the tank it should continue to pick up excess heat and move farther from the allowable curve. Instead of moving away from allowable, the pressure/temperature combination drifted toward it and forty-five minutes later was within limits.

It was the worst kind of anomaly; a discrepancy that corrects itself! Is it the hidden clue to a real problem with the hardware, or just a minor and meaningless aberration in a system so complex that all of its variations may never be understood? If the tank really had a heat leak we should scrub the launch and change it, because otherwise we could not be sure of conducting all the descent engine firings required by the mission. It would delay the launch by about five days.

George Low, NASA's Apollo spacecraft director, came into the SPAN Room himself to confer directly with me and Owen Maynard, NASA's LM Engineering manager. This showed the gravity of the situation, since I had never seen him in SPAN before. Low was a careful, thorough engineer and a decisive manager; he had led the Apollo program's technical rebuilding after the dark days of the Apollo 1 fire. After we briefed him on the helium anomaly, he fixed his steely blue eyes on me and asked for Grumman's recommendation: Did we proceed with the launch or scrub and change the tank? He said I had twenty minutes to let him know.

I spent most of the twenty minutes on the phone with people who could contribute to the decision, including Grumman's corporate chief engineer, Grant Hedrick, who had a sixth sense for pinpointing the cause of technical problems, and the experts at Airesearch, who had searched all prior test records on that tank and found no anomalies. Then I went up to the VIP viewing area behind Mission Control and huddled in a corner of the dimly lit room with Joe Gavin. Summarizing the data and the tradeoffs, I recommended to Gavin that we launch. After a thoughtful pause, he concurred. I hurried back to the SPAN Room and repeated this recommendation to Owen Maynard. When he agreed as well, I called George Low and told him that Grumman officially recommended that NASA proceed with the launch, as we believed the tank to be sound. He asked what Gavin and Hedrick thought. I said they agreed, and Low then said NASA would proceed with the launch. He thanked me for meeting my deadline.

NASA did not mention the issue for the remainder of the mission, but I monitored the helium tank's performance almost continuously whenever it was accessible on the screen.

A Great Flying Machine

Nothing that followed in the actual flight put so much pressure on me personally, and for the most part the flight went well. It was our Grumman support team's first direct experience with astronauts on a real mission, and I found it exciting to have men whom I knew up in space flying our machine. The giant three-stage Saturn 5 booster lifted off on schedule and performed flawlessly except for some Pogo longitudinal vibrations in the S2 stage, placing the spacecraft into exactly the planned Earth orbital altitude. The critical maneuvers of command and service module separation from the spacecraft/LM adapter, and rotation and docking to the LM, went perfectly. Upon command the LM was separated and pulled away by the CSM, while the S4B stage was jettisoned into a lower orbit and burnup in the Earth's atmosphere. After six hours of checking out the CSM and its systems, McDivitt fired the service propulsion system (SPS), and the powerful rocket engine boosted the heavily laden CSM/LM combination into a higher orbit. He sounded relieved that the dormant LM was still there after the force of the first burn. Three additional SPS firings were successfully accomplished, increasing the crew's confidence in the capability of their spacecraft. Following these operations the crew settled down for a meal and sleep; the first Apollo mission on which the three astronauts were allowed to sleep simultaneously. I took advantage of the quiet time and shift change to hand the SPAN duty over to Howard Wright.

I was back to the SPAN Room early the next morning, listening to the crew puffing as they donned their spacesuits to enter the LM. The crew channel went dead. We did not learn until the postflight briefings that Schweickart had suddenly vomited.[6] After some delay he entered the LM and flipped dozens of switches to activate its systems. He commented that the LM was quite noisy, particularly its environmental control system. McDivitt joined him, and after they unpacked the television camera in the LM cabin we watched them on worldwide TV. Our friend McDivitt promptly embarrassed us by pointing out to the world a washer and other bits of manufacturing debris floating through the cabin under zero gravity. It was a chastisement we deserved, and it motivated us to still more stringent efforts to clean the cabin and all closed compartments of the LM during assembly and test.

McDivitt and Schweickart extended the LM's landing gear, which locked smartly into place upon command. They checked out the LM's systems and fired the LM descent engine for more than six minutes at full thrust while in the docked condition, simulating much of the powered descent burn that would be required to bring LM down from lunar orbit for landing. The crew controlled the engine manually and demonstrated digital autopilot attitude control.[7] All LM operations went perfectly, including the performance of the suspect descent helium pressurization system. When McDivitt and Schweickart rejoined Dave Scott in the command module, they felt that their LM would be up to the challenges ahead.

The fourth day in orbit all three astronauts donned their space suits and opened the hatches of both spacecraft. Schweickart and Scott performed spacewalks from the LM and the CM respectively; the former using his backpack[8] for space life support, while Scott and McDivitt's spacesuits were connected to their spacecraft by flexible umbilical hoses. Unknown to us, Schweickart was instructed to take the spacewalk a step at a time and end the exercise immediately if he felt nauseous. Plans for him to float freely in space on his tether were dropped, but he did use the handholds to climb up the front face of the LM ascent stage near the docked connection between Spider and Gumdrop, where he could clearly see Dave Scott standing in Gumdrop's open hatchway. He and Scott photographed each other in their celestial perches like ordinary tourists. During the spacewalk Schweickart went by the call name Red Rover since he was a third spacecraft himself, communicating through the backpack's radio. Although its duration was halved to one hour, the spacewalk and backpack demonstration were completely successful.

The fifth day in orbit was the crucial part of the mission for the LM—the demonstration of LM's flight maneuverability, and its ability to rendezvous in orbit from a far distance. My colleagues and I scrutinized the instrumentation readouts on our consoles carefully as the crew reactivated Spider's systems. Hundreds of pressure, temperature, voltage, current and other measurements located in all the systems were sampled several times a second, giving us detailed real-time information on the LM's health and performance. Any measurement that strayed out of preset normal limits triggered caution and warning alarms in the cockpit and on the ground consoles. With all systems activated, Spider looked good to the crew, to the flight controllers, and to me. Over the net came Flight Director Gene Kranz's crisp voice: "Apollo 9, you're 'go' for LM sep" (lunar module separation).

No longer joined at the head to Gumdrop, Spider cavorted briefly, testing her reaction control system, and then pirouetted slowly before Gumdrop's windows, preening for Dave Scott's inspection. He pronounced her beautiful. After forty-five minutes of maneuvering within 3 miles of Gumdrop, McDivitt fired the descent engine, putting more distance between the two spacecraft. He encountered "chugging" of the descent engine between 10 and 20 percent thrust, which smoothed out completely above 40 percent. Subsequent firings increased the separation distance to over 110 miles, where the pilots could no longer see each other's spacecraft. Spider's crew then separated the ascent from the descent stage while igniting the ascent engine in an orbital simulation of lunar liftoff, and successfully completed orbital rendezvous with Gumdrop, which could be seen from over 45 miles away. Everything worked perfectly except Spider's tracking light, which failed at stage separation. McDivitt and Schweickart spent more than six hours in Spider apart from Gumdrop, exercising all the systems. They proved in flight that the LM could leave the CSM, find its way back to it, and dock safely.[9]

Spider performed so consistently well that I never felt any apprehension as I watched each critical event of the mission click off like clockwork. I could hardly believe that this agile machine, dancing so gracefully through space, was the same crotchety beast with the broken wires and structural cracks that had given us fits for over two years of ground testing. Was our LM design and construction really good after all, or were we just lucky? I was not sure, but thought it was some of both. I was glad we had practiced all those mission simulations. Although the anomalies during the real mission were benign compared with the deviltry concocted by NASA's simulation director, we handled them effectively, and we had an excellent follow-up system in place to assure that the entire list was closed out before the next mission, which would be even more demanding.

After they returned to Gumdrop, the crew closed Spider's docking hatch for the last time and set her free. Ground control radioed a firing signal to Spider's ascent engine to park her in a highly elliptical orbit, and the crew watched her depart until no longer visible. As in the coming lunar missions, Grumman's job was done, even though the mission was still in progress and the crew in space. I joined the rest of my colleagues at the airport and flew back to New York, where four days later I watched Gumdrop's triumphant splashdown in the Atlantic near Puerto Rico. An hour later the crew was safely aboard the USS *Guadalcanal*, grinning broadly and obviously glad to be home safely. At a jubilant ceremony in Washington, Vice President Spiro Agnew awarded NASA Service Medals to Gen. Rip Bolender and Lew Evans to acknowledge the lunar module's successful initiation as a manned space flying machine. The way was now clear for the final rehearsal, a mission that would take the complete Apollo spacecraft around the Moon.[10]

The best accolades to my ears came at the crew's postmission debriefing in Houston. McDivitt and Schweickart were enthusiastic about the LM's performance: "That LM is a great flying machine. And when it's just the ascent stage alone, it's very quick. It snaps to the controls like a fighter plane, or a sports car. It was super to fly!"

LM Brushes the Moon: Apollo 10

"Son of a bitch!"

Apollo 10 lunar module pilot Gene Cernan's startled exclamation over the network snapped me out of the lethargy in which I had been dreamily scanning the LM's readouts on the console monitor. Suddenly there were wild excursions in pitch and yaw, and much reaction control system thruster firing. Snoopy seemed to be throwing a fit, thrashing about in space. Howard Wright was already on the phone with Jack Russell, Grumman's stabilization and control subsystem engineer in nearby Building 45.

"Ask them to be sure the AGS is still in attitude hold," Russell advised immediately.

Wright and I passed this word on to Scott Simpkinson, NASA's senior person in the SPAN Room. We could hear the crew's heavy breathing over their open mike on Snoopy as Comdr. Tom Stafford struggled to regain control. Stafford jettisoned the LM descent stage, the next planned activity before igniting the ascent engine to leave low lunar orbit and rendezvous with John

Young in Charlie Brown. Despite Snoopy's being over thirty degrees off its proper flight attitude, this critical function was executed flawlessly—explosive bolts and nuts, guillotine cutter, deadface connector fired in unison, and the descent stage fell away stably as Stafford ignited his RCS thrusters. The released ascent stage continued to thrash spasmodically about all three axes. A warning light said they were approaching gimbal lock of the guidance platform. Stafford took over manually and worked the attitude control switches, and Snoopy calmed down.

Playing back the last few minutes' data, the flight controllers determined that the AGS had been mistakenly switched to automatic mode while the crew was correcting a minor rate gyro disturbance, and Snoopy's tantrum was the result. Although AGS was normally a backup to the primary navigation and guidance system, to be used only if PNGS failed, for Apollo 10 it was being used to control the ascent into rendezvous orbit to demonstrate its capability in flight. In the automatic mode it searched for and locked onto the command and service modules, producing the unwanted attitude gyrations. Stafford's switching had already restored it to the correct "attitude hold" setting.

Over the net I heard CapCom Charlie Duke tell the crew they had corrected an improper AGS switch setting that had caused the disturbance, and everything on Snoopy looked good. They were cleared to fire the ascent engine for rendezvous orbit insertion. The whole unsettling episode had taken about three minutes.

The ascent engine ignited smoothly, and Snoopy ascended from skimming the forbidding mountains of the Moon at fifty thousand feet as its orbital velocity increased. Soon we would know if the orbital mechanics, rendezvous procedures and communications that had been demonstrated in countless simulations, and in Earth orbit on Apollo 9, would work as well in close proximity to that enigmatic gray eminence, the Moon. Its uneven mass concentrations (mascons) could perturb the analysts' orbital calculations; its lack of shielding atmosphere left Apollo's radios open to solar radiation interference. These concerns, together with the desire for close reconnaissance of the tentatively selected first lunar landing site, were NASA's basis for performing the Apollo 10 low-altitude lunar-orbit mission, instead of going directly for a landing.

Stafford and Cernan enjoyed their smooth upward ride, and commented that they could see the Moon's features receding away from them. At the end of the ascent burn, they were in the correct attitude and flight path for rendezvous, 48 miles from Charlie Brown, on whom they had radar lock and visual sighting. Snoopy's crew had first seen Charlie Brown from 100 miles away, and Young had seen them through his sextant at 155 miles. Steady communications with each other and Houston kept both crews aware of what was happening. Rendezvous closure and docking were routinely successful, as Stafford skillfully positioned the skittish lightened Snoopy close to Charlie Brown, and Young firmly thrust his probe into Snoopy's drogue and was rewarded by the reassuring snap of twelve capture latches engaging.[11] Lunar-orbit rendezvous, so long debated and studied with apprehension, proved to be a "piece of cake."

With the crew reunited in Charlie Brown, Snoopy was jettisoned, and its ascent engine was again commanded to fire, placing the LM ascent stage into a long, elliptical orbit around the Sun. Wright and I packed up and left the SPAN Room for the airport, our job completed. Except for those three minutes of excitement, we had little to do in SPAN because Snoopy performed so well. The descent engine showed none of the chugging that alarmed McDivitt on Apollo 9, probably because of some minor modifications we had made to the engine controls. The Saturn boost on Apollo 10 was a hard one, with rough Pogo vibrations on all three stages so strong that the crew worried if the spacecraft had held together, but neither Snoopy nor Charlie Brown showed any ill effects.

Three days later the Apollo 10 crew was safely aboard the USS *Princeton* in the Pacific, their mission accomplished with finesse. The final hurdle had been cleared, and Apollo was ready to land on the Moon. Looking up at that smiling white orb on a mild late spring night, I felt a profound sense of wonder. It was really going to happen—men would walk on the Moon. Recalling all the years of work, failures, and frustrations, it was hard to believe that our dogged, fumbling efforts were close to achieving their goal. In three spaceflights our LM had performed better than I believed possible—nothing like the problem-ridden ground tests I remembered from S/CAT. Each LM was markedly improved over its predecessor, as the Grumman Engineering, Manufacturing, and Test teams became more skillful at correcting problems and devising better ways to do things. Very soon we would see if LM-5, the next in line, would be up to all the challenges that landing and taking off from the Moon's unknown surface would entail.

15

One Giant Leap for Mankind

Apollo 11

On 16 July 1969 I settled into my familiar desk in the SPAN Room and watched and listened, on the TV monitors and over the net, the awesome spectacle of the 330-feet-tall, six-million-pound monster called Saturn 5 awakening with a rumbling roar of bright orange flame and black smoke. Shaking the white, crumbling ice sheath off its flanks like a gigantic fire-breathing dragon shedding its skin, and slowly shuddering upward, it gathered speed as its huge bulk cleared the launch tower. The thunderous roar hit the crowds on the beach and in the viewing stands a few seconds later, seeming to come up from the vibrating ground below as well as from the sky above. In a crackling, roaring jumble of sound the giant began to look smaller and more distant as, ever accelerating, it gained the speed that would place it far out of sight and into orbit.

As command was switched from Kennedy Space Center to Houston, a reassuring stream of "Go" and "Looking good" flowed from the booster flight controllers, and the large altitude-velocity plot on the wall in Mission Control showed a steady progression of real data points moving up the predicted line. After two minutes and forty seconds the mighty S-1C stage completed its burn and dropped away, taking its five outsized, 1.5-million-pound-thrust rocket engines with it. We saw the stage separation in the TV image from the range tracking camera. Six and a half minutes later the S-2 stage, which burned hydrogen and oxygen with a brilliant white flame, shut down and was jettisoned, leaving the single rocket engine of the S-4B stage to compete the burn to Earth orbit. When later reignited, the S-4B provided the remainder of the twenty-six-thousand-mile-per-hour velocity required to escape Earth's gravity and reach the Moon. Verifying the fixes made since Apollo 10, the crew reported almost no Pogo oscillations from the Saturn stages.

The Real Thing

After all the simulations and preparatory missions, it was hard to believe that this was the real thing; they were going to make our LM do everything it had been designed to do this time. I prayed that everything would work right.

Coasting silently in an escape trajectory toward the Moon, Columbia separated from the spacecraft/LM adapter, the hollow truncated cone that housed the LM during liftoff. She turned around and docked with Eagle and jettisoned the SLA. It was Eagle and Columbia, joined head to head and sailing to the Moon. The next two days passed uneventfully, with the spacecraft on target and looking good, the LM still quiescent.

Occasionally I walked down the hall to the VIP viewing area, where I could watch the activities on the main floor of the Mission Operations Control Room. This busy room, packed with flight controllers and their consoles, had become a familiar sight to people the world over, as the TV coverage frequently showed the action and zoomed in on some of the key players, with an all-knowing commentator piously intoning to the world what they were probably talking about, based on the mission problems of the minute. Speakers in the VIP area played the NASA public information channel coverage of the flight, which included all the open channels used by the flight director, CapCom, and the astronauts. The VIP area was usually almost empty during the noncritical mission times when I could drop in, but Joe Gavin and Lew Evans were liable to be there at any hour, so it was a good place to meet informally with them.

Grumman's president Lew Evans enthusiastically supported the company's ambition to play a major role in the Apollo program. He took his role as Grumman's leader very seriously. Alone among the Apollo contractor presidents he made a point of being on public display in the VIP area when LM was active in a mission. He solemnly said he did it so he could be there to take our lumps if anything went wrong with the LM—or to receive congratulations if all went well, as the dour look gave way to his infectious Welsh grin. A nontechnical lawyer and businessman, he had implicit faith that Gavin and I would work our engineering magic to keep Grumman from disgrace.

His resolve did not go unnoticed by the NASA leadership. NASA Associate Administrator for Manned Spaceflight George Mueller, Apollo Program Director Gen. Sam Phillips, Apollo Spacecraft Director George Low, and Manned Spacecraft Center Director Bob Gilruth all made a point of dropping by to see Evans during the missions and occasionally waved or nodded to him from their consoles in Mission Control through the VIP area's glass partitions.

The outbound voyage of Apollo 11 unfolded smoothly. The real mission, like Apollo 10 before it, was much quieter than a typical mission simulation, with few anomalies and no hard failures. Two and a half days after Earth launch, Aldrin entered the LM and checked out its systems. We eagerly scanned the wealth of data that came flooding down from Eagle onto our screens and were delighted to see that she had come through launch and translunar coast in good shape. Just after Apollo 11 passed behind the Moon, out of touch with Earth, the service propulsion engine was fired for six minutes to send the spacecraft into lunar orbit. The crew performed navigation checks, then a second short burn to circularize the lunar orbit at about sixty thousand feet above the surface. Armstrong and Aldrin entered Eagle and performed a thorough systems checkout. There were no anomalies. They shut her down again and returned to the CM for a nine-hour eat-and-rest period—the last break before their descent to the Moon.

During quiet mission periods I rotated duty at the Grumman desk in SPAN with my deputy Howard Wright and Chief Subsystem Engineer Bob Carbee in eight-hour shifts. When not on duty we could continue to watch the action from the VIP area or the Engineering support floor in Building 45 or return to our nearby motel for food and rest. Even my motel room had the NASA public affairs channel and, of course, a TV. For the landing all three of us wanted to be in the SPAN Room, and NASA, though concerned that overcrowding might detract from our effectiveness, reluctantly concurred.

After a fitful rest Armstrong and Aldrin entered Eagle and activated her systems again. Leaving Collins alone to tend Columbia, the hatches between the two spacecraft were closed, and with a "pop" the vehicles separated.

"The Eagle has wings!" Armstrong called out exultantly, as he maneuvered his landing craft for the first time, pirouetting in front of Columbia's windows so Collins could visually inspect Eagle. While behind the Moon, a brief burst from Eagle's descent engine dropped her below and behind Columbia, heading for an orbital low altitude (perilune) only nine thousand feet above the surface, from which point a continuous twelve-minute burn of the variable thrust descent engine would complete the landing.

When Houston again acquired Eagle as she emerged from behind the Moon, less than half an hour remained to touchdown. I thought I would explode from pent-up nervous energy, and yet I was fascinated as one after another the events required in the flight plan flashed by. LM's rendezvous and landing radars were both on and operating; everything else was looking good. After the leisurely translunar coast phase, mission events whirred by as though on fast forward. Shortly after the final descent engine burn began, several program alarms occurred in rapid succession. In each case the flight controllers assured the crew they could reset the warning indicator and press on. One of the alarms was due to an improper switch position in LM; two others were momentary computer memory buffer overloads with data coming in from the radars. These were resolved by NASA so quickly that Grumman support input was neither possible nor necessary. We held our breaths as Armstrong said he was taking over manually to avoid boulders he could see at the programmed landing point, listening as Aldrin called out the few feet

Supporting Apollo 11 in the Spacecraft Analysis Room. *At desks, left to right:* Tom Kelly, Owen Maynard, Dale Myers, and George Merrick. (Courtesy NASA)

remaining of altitude and velocity. Then they were down, engine off, and we heard Armstrong's historic words: "Houston, Tranquility Base here. The Eagle has landed!"

Cheers resounded in Mission Control and the SPAN Room, but Flight Director Gene Kranz quickly restored order, reminding us of the heavy post-landing activity ahead, including immediate preparation for liftoff in case lunar stay had to be curtailed for any reason. Carbee, Wright, and I embraced each other in joy and relief that the landing was safely accomplished. Seven years we had worked for this moment; how marvelous to experience it! I saw tears streaming down Carbee's cheeks and felt a lump in my own throat as well.

Our joy was short lived. Within a minute after landing my phone rang. It was Manning Dandridge over in Building 45, pointing out to me that the pressure and temperature were rising in a section of fuel line between the descent helium heat exchanger and the fuel control valve on the descent rocket engine. This was quickly picked up on the net by the LM propulsion flight controller, Jim Hannigan. Dandridge and George Pinter (our supercritical helium

Celebrating the Apollo 11 Moon landing. *Left to right:* Arnold Whitaker, Bob Carbee, Lew Evans, Tom Kelly, Jim Leather, Frank Canning, and John Coursen. (Courtesy Northrop/Grumman Corporation)

expert) concluded that after engine shutdown, cold helium froze the fuel remaining in the heat exchanger, trapping liquid fuel between the frozen fuel slug and the closed engine fuel valve. Heat soaking back from the hot engine parts after the long burn increased the pressure and temperature of the trapped fuel.

Hannigan left his console in the MOCR and joined us in SPAN. We put Dandridge and the Building 45 people on the speaker phone; Gerry Elverum, the chief engineer of the descent engine for STL was with them. As the temperature exceeded three hundred degrees Fahrenheit we became very concerned. The rocket fuel was a mixture of two forms of hydrazine, and at temperatures above four hundred degrees it became unstable. Extrapolating the rate of temperature rise suggested it would exceed that limit in ten minutes. We all felt that the consequences of an explosion, even of the relatively small amount of fuel remaining in that short section of line, was unpredictable and unacceptable. Phone calls to other rocket experts at Aerojet and Rocketdyne confirmed this view.

Dandridge suggested the possibility of "burping" the descent engine by commanding it to fire, then immediately shutting down, to momentarily crack open the engine fuel valve to relieve the pressure and temperature. George Low and Gene Kranz then joined us in SPAN and asked a barrage of pertinent questions, especially concerned whether we could cause the LM to tip over.

We decided that a momentary flick of the manual firing button was the best procedure. Since the trapped volume of fuel was small, even rapid opening and reclosing of the valve would provide major relief in pressure and temperature. Temperature was now over 350 degrees; we had only a few minutes left to act.

Low decided to do it and returned to Mission Control to explain the procedure and reasoning to CapCom Charlie Duke for transmission to the crew. Before he could place the call, however, nature took over and solved the problem for us. Heat soaking back from the engine melted the fuel ice plug in the heat exchanger and the pressure abruptly dropped to a low value. We looked at the screen in amazement for a few seconds, then broke into smiles and cheers of relief. I collapsed back into my chair and realized that I was drenched with sweat. What a welcome to the Moon! To this day, when asked how I felt right after the first LM landed on the Moon, I say that for ten minutes following touchdown I was too busy and worried to even know they were there.

The crew was working their postlanding checklist and, finding all in order, received clearance to stay on the Moon. They had no inkling of the propulsion system drama that had just played out six feet beneath them. The only problem they reported was that the mission timer had stopped and would not reset—strictly a nuisance item, although we would have to find the cause and fix it before the next mission.

I also carefully looked over the LM data, but I could find nothing else wrong. At last I could believe the landing was successful! Soon, in addition to relief, I felt a growing curiosity about the details of the landing site. Before they proceeded with the required practice liftoff countdown, the crew spent a few minutes at Eagle's windows taking a look at their new surroundings. I imagined the scene at Tranquility Base as the crew described the strange other world outside. They saw a flat plain, pockmarked with craters and studded with rocks and boulders of diverse sizes, and dry, dusty-looking soil. Colors varied from white to shades of gray and brown, depending on the Sun's angle to the object, and the sky was jet black with thousands of stars showing, graced by the spectacular blue-and-white bauble that was Mother Earth. The brief reports from the crew mostly stuck strictly to factual descriptions but occasionally ventured a characterization, such as "awesome beauty" or "magnificent desolation." They returned to the practice countdown, which verified all LM systems and configured them for liftoff. Then they settled down in

Eagle for the scheduled activities of eating and sleeping. My colleagues and I took advantage of the quiet period to eat in the MCC cafeteria, where we exchanged congratulations with Bob Gilruth and George Low. We then had a rest period of our own at our nearby motel.

I slipped into a sound sleep when the telephone woke me. It was Bob Carbee in SPAN with word that the crew was getting ready to come out onto the Moon early, three and a half hours ahead of the flight plan. How could they sleep when they had just landed on a new world? Excitement ran high in the SPAN Room. I listened eagerly to the crew preparing for egress, donning spacesuits and checking out and donning the backpacks. The crew provided running commentary of their activities. Despite their care in moving about the cabin, Armstrong accidentally broke off the button of a circuit breaker, which protected the ascent engine arming function, with his bulky backpack. I asked our Grumman support people in Building 45 and at Bethpage to work out an alternate procedure for arming the engine without that circuit breaker. There were several different paths for executing this critical function, so I was confident that we could find a way around it.[1] The astronauts depressurized the LM cabin, opened the forward hatch, and Armstrong cautiously backed out onto the "front porch" platform, with Aldrin giving him helpful steering instructions from inside. With a black-and-white TV camera on LM and the whole world watching, he climbed down the ladder and set foot on the Moon: "That's one small step for [a] man; one giant leap for mankind."[2]

Armstrong experimented with various gaits for moving about on the powdery charcoal-like surface, settling on a lengthened walking stride. Aldrin joined him, and for the next two and a half hours they explored the area. They took the TV camera outside and mounted it on a stand, sharing the sights and their activities with the world. They planted the American flag on the surface and answered a telephone call from President Nixon. Opening the scientific equipment compartment on the LM descent stage, they deployed seismic sensors, a solar wind experiment, and a laser reflectometer. They took photographs and gathered lunar soil and rock samples. So little time, and so much to see and do!

I was fascinated to see how well the LM had weathered the landing. Touchdown was gentle, so the energy-absorbing landing-gear struts were hardly compressed at all and the footpads had only sunk about two inches into the dust. The silver, black, and gold LM was the only spot of color in an otherwise gray world. The astronauts adapted well to the Moon's gravity, and with a little practice they were confidently hopping about the area, investigating interesting craters, boulders, and rocks. Even in the ghostly black-and-white TV images the strange timeless beauty of the barren moonscape was evident. How privileged I felt to be playing a part in such an adventure.

Things went so well that Mission Control allowed the explorers to stay outside for an extra fifteen minutes. All too quickly time was up and they were carrying their full sample containers up onto Eagle's front porch and returning to the cabin. The gray lunar dust clung to everything—the crew stamped their feet, clapped their hands, and brushed off their legs and the sample boxes in an attempt to leave most of the dirt outside.

The crew repressurized the LM, doffed their backpacks, and stowed the lunar sample containers, cameras, and other equipment. When the cabin was shipshape, they depressurized it and opened the forward hatch for the last time in order to throw out their backpacks and other items no longer needed. Repressurizing once more, they then removed their helmets and gloves, settling down for about six hours of eating and sleeping before preparation for liftoff. This time they really slept, and so did I back in the motel. But I kept waking up and worrying about a micrometeorite strike or window failure causing a sudden depressurization, leaving the crew gasping for breath in an airless cabin. I knew the odds were overwhelmingly against this, but that did not stop my worrying.

Although the descent and landing was a greater challenge to piloting skills and involved the unknowns of the lunar surface, I thought the liftoff from the Moon was the most daring event in the LM-active portion of the mission. The challenge of performing launch operations with only two men, a quarter of a million miles away from Earth, seemed impossible when compared with the eight thousand people and weeks of preparation required to perform the same function from Cape Kennedy. It was a measure of how successful we had been in keeping the LM simple in design and operation, and of the competence of the astronauts and the thoroughness of their training, that lunar surface launch was not only possible but even seemed a natural part of the mission.

The liftoff did not allow for introspection or corrective action—either it worked or it did not. At the same instant as the ascent rocket engine's valves opened and ignition occurred, a string of explosive charges shattered the bolts and the nuts holding the ascent and descent stages together, a guillotine cut a four-inch-diameter umbilical wire and tubing bundle connecting the stages, and a deadface connector removed power from the severed wires in the bundle. If the engine did not ignite or the stages failed to separate, the crew would be doomed to remain on the Moon, dying before the watching world, a fate too horrible for me to contemplate.

For the next seven minutes the crew stood a foot in front of the ascent rocket engine as it fired with only a thin aluminum can between them and the combustion chamber. On Earth, rocket firings were conducted from the safety of a sandbagged and partially buried blockhouse behind thick walls of reinforced concrete. The astronauts were courageous, yet confident in the integrity of their spacecraft and all who designed and built them. It was a trust that weighed upon me heavily.

Tension mounted in Mission Control as the crew went through the prelaunch countdown, and then bang! it happened, just as it was supposed to. Liftoff was smooth—from the windows the pilots could see a shower of torn insulation and Aldrin glimpsed the American flag being blown down. There was little feeling of acceleration, but when the ascent stage pitched over to its climbing trajectory the windows faced the surface and they could see that they were going faster and faster. The crew called out some landmarks they recognized on the way up. It was like a high-speed elevator ride, with little noise and vibration despite the proximity of the rocket engine and no sensation of speed if you did not look out the window.[3] The engine shut down on schedule, placing them in a nearly perfect orbit for rendezvous.

I was much less worried about the rendezvous and docking maneuvers. These operations had been practiced many times on Earth in high-fidelity simulators and verified in flight on Apollos 9 and 10. Moreover, it was a non–time critical "slow motion" operation with several backup paths using independent equipment. Collins in Columbia had a checklist of eighteen different procedures to use if the primary method failed.[4] The probe-and-drogue docking mechanism, with its capture latches firmly holding the two spacecraft together, was the only element for which there was no direct backup, but it was a relatively simple, rugged design with broad dimensional tolerances.

On Apollo 11 the rendezvous maneuvers were skillfully executed as planned. The reunion was accomplished with the primary method using the rendezvous radar, right in the center of the tolerance band. The hard docking was a bit more sporty—Collins had to manually counter a sudden yaw disturbance just before the latches snapped closed. He readily removed the probe and drogue for the lunar explorers to rejoin him, relieving his recurring worry that the drogue would stick in place, blocking access from the LM.[5] Everything worked, and the three were joyously reunited, none more relieved than Collins, who had agonized during his stay in lunar orbit over how he would deal with returning alone to Earth if Eagle and her crew did not make it back.[6] They transferred equipment, cameras, and the lunar samples into Columbia, leaving the probe and drogue, their functions completed, in Eagle for disposal.

With the crew safely aboard Columbia, Eagle was jettisoned, moving away proudly with her attitude control thrusters firing, into a low orbit that would eventually crash on the Moon. Howard Wright and I took off our headsets and looked at Eagle's data on the screen for the last time. Our part of the mission was completed; there was nothing more that Grumman could do to help Apollo 11. We packed up our reference books and papers, accepted congratulations from our NASA and North American colleagues in the SPAN Room on the LM's sterling performance, and headed for the airport and home.

Of course there was no celebration until the crew was safely aboard the aircraft carrier USS *Hornet*. Two and a half days after Eagle was jettisoned, I watched the orange-and-white-striped parachutes lazily floating down toward the blue Pacific, surrounded by my Grumman friends and associates in the Mission Support Center in Bethpage. We broke into a roar of triumph as the TV showed the conical command module splashing into the water and the parachutes deflating and gently settling down beside it. From the screen we saw American flags and cigars broken out by the flight controllers in Mission Control in Houston—we made do with arm waving and cheers. It was a moment of elation, of heart-pounding thrills usually reserved for athletes or warriors, for a successful team effort, a group victory achieved against long odds. In reaching the Moon and safely returning, the hundreds of thousands of ordinary Americans who had labored for most of a decade on the Apollo program briefly shared this privileged feeling.

16

Great Balls of Fire!

Apollo 12

Pete Conrad was unlike most of the other astronauts. Voluble, excitable, enthusiastic, and totally approachable, he had none of the "right stuff" reserve displayed by some of NASA's superstars. Nor, like many of his colleagues, did his appearance suggest that Hollywood's central casting had chosen him for the role of spaceflight hero. Short, balding, and with widely spaced upper front teeth, he looked more like a jockey than an astronaut, but he was a seasoned navy carrier and test pilot, and a proven veteran of the Gemini program. He was Gordon Cooper's copilot on Gemini 5, an eight-day endurance test in Earth orbit, and he commanded Gemini 11, which made the first one-orbit rendezvous with an Agena target vehicle (to which copilot Dick Gordon made a risky spacewalk to attach an experimental tether). Beneath his easygoing attitude and wisecracking lay competence, experience, and sound judgment.

I met Conrad early in the LM program, at our first mockup review, and worked with him for the next several months as he helped us develop the LM flight stations and lunar surface access provisions. He was fun to work with—his antics as he swung suspended in our awkward Peter Pan rig were slapstick comedy at its best. But he also convinced us that a platform and ladder were needed on the LM's forward landing gear strut, instead of the block-and-tackle arrangement we had proposed. He was helpful in improving the flight station arrangements and the controls and displays, and in gaining acceptance of the other astronauts for the evolving design. Among other LM design details Conrad helped develop were the rectangular forward hatch, the overhead docking window and electroluminescent lighting for the displays and instruments. He was so enthusiastic about the virtues of electroluminescent lighting that he sold it to Joe Shea and Bob Gilruth for the command module as well.

After our M-5 mockup review Conrad went off on the Gemini program and I did not see much of him until he and Alan Bean began coming to Bethpage for some of the LM-3 tests, in their role as backup crew for Apollo 9. He greeted me and my colleagues like long-lost buddies, and we picked up our friendship immediately, despite almost three years' absence. Pete Conrad and Al Bean were not only dedicated and diligent in performing tests whenever our around-the-clock schedule required but also good-humored about it.

When Conrad, Al Bean, and Dick Gordon were assigned as prime crew for Apollo 12, Conrad and Bean were often in Bethpage testing LM-6, to which they gave the mission call-name Intrepid. Bean, LM pilot, was serious and studious and delved deeply into detail until he thoroughly understood how all the LM systems worked and what could go wrong with them. He would spend all night following up on one of Conrad's offhand comments during a test until he had a complete explanation that Pete would buy. I thought they made a great team, and I sought their suggestions on how we could improve LM design and procedures. After LM-6 was delivered to KSC, I did not see them again until the flight readiness review, where they said they were satisfied that Intrepid was in good shape and ready to fly. They, who would be risking their lives in my spacecraft, urged me, who would be watching safely from the ground, not to worry; everything with LM would be just fine. They surely were not lacking in confidence!

Thus it really got my attention when Pete Conrad coolly called out to Houston a few minutes after the Saturn 5 lifted Apollo 12 into low clouds at KSC, "Okay, we just lost the platform, gang. I don't know what happened here; we had everything in the world drop out."

A look at the SPAN video monitors confirmed what the crew and the flight controllers were seeing: command module Yankee Clipper showed a master alarm, its platform was tumbling, and all instrumentation readouts showed gibberish. The crew saw all lights lit on the caution and warning panel, something that had never been served up in the most imaginative ground simulation. Al Bean was puzzled, because although the readouts indicated the entire electrical system was out, current meters on the panel showed the spacecraft was still drawing power, but at a lower than normal rate. Conrad calmly read the long list of warning lights to Houston and waited for advice. Meanwhile, the Saturn continued its ascent unperturbed, under the control of its inertial measurement unit, which was independent of the CM's guidance platform.

Flight Controller John Aaron looked at the meaningless pattern of numbers coming from the CM's instrumentation and remembered seeing it before, when he was watching a KSC launch preparation test a year earlier. He advised Flight Director Gerry Griffin to ask the crew to switch the signal conditioning equipment to the auxiliary position. Neither CapCom Jerry Carr nor Pete Conrad recognized the obscure command, but Al Bean found the proper switch and moved it, and instrumentation readouts were instantly restored. With live data to study, the crew and ground controllers saw no residual damage, except that the fuel cells had disconnected from the electrical buses, leaving the spacecraft without its primary power source,[1] and the guidance platform had tumbled. Immediately following first-stage separation and S-2 stage ignition, Bean recycled the electrical bus circuit breakers and the fuel cells came back on line. The platform could be realigned when they were safely in orbit, which is where they were headed.

"I'm not sure we didn't get hit by lightning," Conrad told Houston. He was right; later analysis of data and automatic cameras at the launch pad showed that two lightning bolts had struck Apollo 12, at thirty-six and fifty-two seconds after liftoff. The electrical surge from the first strike caused the CM systems to shut down, and the second knocked out the guidance platform. The long column of ionized gases from the Saturn rocket exhaust had acted like a giant lightning rod, providing an attractive path to ground from the dark clouds overhead.[2]

When Apollo 12 was in Earth orbit, NASA and North American had an hour and a half to determine whether it was safe to fly to the Moon. As Dick Gordon was taking star sightings to realign the platform, Mission Control determined that all CM system data looked good except for a few instrumentation sensors, which had apparently been burned out. How could they quickly determine whether Yankee Clipper's systems were all right? Again flight controller John Aaron made a key contribution, proposing that they perform the lunar orbit injection (LOI) checklist, the most demanding flight checkout the CM would have to pass, and if it passed that it was okay to fly. Flight Operations Director Chris Kraft told them to proceed with the checklist, but reserved judgment on whether this would clear Apollo 12 for flight until he consulted with other senior NASA managers. Gilruth, Low, McDivitt,[3] and Petrone approved; their one reservation, that the lightning might have knocked out the pyrotechnic systems on the recovery parachutes, was rendered moot because going to the Moon would not change the crew's final predicament if that were the case. The flight crew and ground controllers successfully performed the LOI checklist, and Apollo 12 was cleared to go to the Moon.[4]

Through all this turmoil the LM Intrepid was snugly nestled inside the protective conical structure of the spacecraft/LM adapter, where it was when the lightning had struck. Inactive and inert, I did not think the LM had sustained any damage, since the major lightning current would travel through the outer skin of the SLA. However, within sensitive electronic microcircuits there was always the possibility of inducing damaging secondary currents in the tiny etched conductors and components, whenever a strong current surge and changing magnetic field was nearby. While LM was inaccessible, there was not much we could do except worry. I asked our section heads to review all electrical and electronic components with their suppliers, to identify those most susceptible to lightning-induced damage. If we could verify the integrity of some of the most susceptible items, we would gain confidence in the LM's overall status.

Half way to the Moon, the crew entered the LM and performed housekeeping and communications checks. Everything seemed all right, but little had been tested. Not until they were in lunar orbit, only six and a half hours before planned touchdown, did Conrad and Bean enter the LM fully suited, activate all systems, and put Intrepid through the extensive powered descent initiation (PDI) checklist. With all eyes in Houston and Bethpage on the monitors, no discrepancies were seen. Our spindly, fragile-looking Intrepid had survived the great balls of fire. I felt greatly relieved, even though I had expected this outcome and confidently predicted it to my associates.

Apollo 11 had landed four miles from its targeted landing point, and for the first few hours Houston and the astronauts were trying to determine exactly where on the Moon they were. Exasperated, Gen. Sam Phillips, Apollo program director, demanded a pinpoint landing for the next mission. To make the requirement clearly visible, NASA decided that the next mission, to the Ocean of Storms, would include a walk to the unmanned Surveyor 3 spacecraft that had landed there in April 1967. To achieve pinpoint landing capability would require inventing new guidance techniques beyond those available for Apollo 11. Highly accurate landings were essential to efficient lunar exploration because they allowed scientists and geologists to plan and discuss with the astronauts, in detail and in advance, the exact route, objectives, and techniques to be used during each excursion onto the lunar surface. Without knowing their exact surface position, much valuable time on the Moon would be wasted in ad hoc revision of the geological exploration traverses and in orienting the explorers to alternate landmarks.

A young NASA mathematician named Emil Schiesser made the crucial breakthrough. He devised an elegantly simple scheme that used the Doppler shift pattern of radio frequency communications from the LM.[5] By comparing the predicted Doppler pattern with the actual observations, ground control's computers could calculate the deviation in the LM's real trajectory from the targeted preprogrammed flight path. NASA flight controllers had already been using this technique to analyze translunar flight trajectories. In meetings of Bill Tindall's mission planning analysis group, a brilliant technique was devised to get the flight path correction information into the LM's computer. The computer would be told the target landing point had moved by the amount necessary to cancel out the deviation between the real and the predicted trajectories. This required the pilot to enter only a single number into the LM's data entry keyboard.

LM-powered descent on Apollo 12 went beautifully. The computer overload problem that had produced heart-stopping program alarms for Armstrong and Aldrin had been corrected simply by changing software instructions to no longer require the computer to keep updating and storing rendezvous radar data with the CSM's position when the landing radar was operating during powered descent. When Intrepid pitched forward for the final descent at seven thousand feet, Conrad anxiously scanned the lunar landscape below, then gave an exultant war whoop: "Hey, there it is. There it is! Son of a gun, right down the middle of the road!"

He recognized the Snowman crater, on whose rim Surveyor 3 rested, far ahead in a sea of other craters. At an altitude of four hundred feet he took control manually and skillfully landed Intrepid close to Snowman's rim, about six hundred feet from Surveyor 3. He performed the final one hundred feet of descent primarily on instruments, since his view of the surface was largely obscured by dust kicked up by the descent engine's exhaust.[6] When the blue lunar surface contact light went on, Conrad shut the engine down and Intrepid thumped firmly onto the ground.[7] They had achieved a perfect pinpoint landing. There was no postlanding excitement as on Apollo 11, since a simple procedural change had solved the fuel-line pressure buildup problem. (We delayed venting the descent propellant tanks after landing, giving the descent engine time to cool off.)

Conrad and Bean were on the Moon for thirty-two hours and performed two moonwalks totaling seven and three-quarter hours. They made the first deployment of the Apollo lunar surface experiments package (ALSEP), an expanded array of scientific instruments that included a seismometer, a magnetometer, an atmospheric particle sensor and a central transmitting station to relay ALSEP's data to Earth. ALSEP was designed to gather and transmit data for five years, and the instruments were so sensitive that the seismometer detected the astronauts' footsteps as soon as it was turned on. We had modified LM-6's scientific equipment bay to accommodate ALSEP and made special external mounting provisions for the radioactive thermoelectric generator (RTG). It used a radioisotope that would provide power to the ALSEP even during lunar night. Because of safety concerns in event of a launch failure, the power source was encased in ablative insulation capable of reentering the Earth's atmosphere.

On their second excursion Conrad and Bean walked to Surveyor, inspected and photographed it, and took back samples to show the effect of two and a half years of lunar "weathering" on various materials. They also gathered rock and soil samples from the rim of Snowman crater and other craters in the vicinity, packing more than seventy-five pounds of samples for return to Earth. They kept up a running commentary on what they were seeing and doing, including many excited cries of awe and wonder. It was a pleasure to share their delight in the new world they were exploring.

As the time for lunar liftoff approached, tension mounted both aboard Intrepid and in Houston. Conrad told Bean not to worry because if the ascent engine did not work, they would become "the first permanent monument to the space program."[8] In SPAN I worried more about the rapid sequence of explosive-device firings that had to take place to separate the ascent and descent stages. All the pyrotechnics were fired by squibs of very thin wire—could the lightning strikes at liftoff have induced currents that burned them out? There was no way to tell. A positive sign was that all prior explosive devices had worked as commanded to deploy the landing gear and to pressurize the ascent and descent propellant tanks and the reaction control system. I just had to hope that the remaining devices were also unimpaired.

The best thing about lunar liftoff was that you knew in an instant if it was successful. And it was, as the ascent stage separated smartly from the descent launch platform and rapidly accelerated toward lunar orbit, Yankee Clipper, and the waiting Dick Gordon. Rendezvous was again a majestic, slow-motion maneuver, precisely executed, and the spacecraft docked and latched together without problems. We heard the joyous cries as the long-separated shipmates embraced and congratulated one another on their good fortune. Apollo 12's crew was expressive and demonstrative, and they sometimes forgot to turn their cabin microphone off, so we in Houston could briefly share in their joy.

Soon the LM's part of the mission was over, and the crew was Earthbound aboard Yankee Clipper. My colleagues and I returned to Bethpage and watched the splashdown and aircraft carrier recovery in the Mission Support Room. I savored the Apollo 12 experience—although there had been plenty to worry about, nothing had gone wrong. Except for failure of the portable TV camera, which Al Bean had burned out by pointing it at the Sun early in the first moonwalk, every mission objective had been met or exceeded. NASA was impressively able to handle unforeseen emergencies like the lightning strikes, and the entire mission support team, NASA and contractors, was growing more efficient and professional. We were learning from each mission and making many improvements in the way we operated. The number of LM anomalies during the mission was lower than on any previous mission—a positive trend for each mission thus far.

Apollo 12 shifted the goal from landing on the Moon and surviving to purposefully exploring the Moon. Clearly the Apollo program was not just a stunt, as some detractors had charged, but a serious and unique opportunity to answer scientific questions that had long puzzled mankind. Where did the Moon come from? How was its origin related to Earth's? How old was the Moon, and what was its geologic composition and history? We would soon have some answers to these and other cosmic questions.

Swept forward on a rising tide of optimism, I decided not even to cover the next mission in Houston but to leave it in the capable hands of my colleagues while I stayed up at MIT in the Sloan Fellows program.

Even at the great distance from Moon to Earth, the irrepressible, bubbling personality of Pete Conrad and his crew had made Apollo 12 unique—a highly successful and productive mission that was also an adventure and a privilege shared by all those who supported it.

17

Rescue in Space

Apollo 13

I groped for the ringing bedside telephone in the midnight darkness. Knocking the receiver to the floor, I groaned as I snapped on the light, shattering for good any pretense of sleep. I felt my wife stir protestingly beside me as I fumbled to retrieve the receiver, trying not to wake her.

"Tom, have you heard the news?" asked the crisp, worried voice of my Grumman colleague Howard Wright.

Negative grunt.

"Well, turn on your radio—Apollo's in trouble. There's been an explosion or something. The company's chartering a plane to fly us to Bethpage. Meet me at the general aviation terminal at Logan Airport at 1:30."

"Are they alive?"

"Yes, but they're in real trouble."

They were in trouble all right, and would probably have to use the lunar module as a lifeboat. There wasn't time for Wright to tell me more than that. He promised to fill me in on the details at the airport.

Joan looked at me wide-eyed with concern.

"It's Apollo," I told her. "They want me to get down to Bethpage right away." We were living in the Boston area on a temporary one-year assignment while I attended the Sloan Fellows management program at MIT. After seven years of total dedication to designing and building LM, and following the successful first lunar landing of Apollo 11, the company decided that I could use a change and allowed me to compete for a Sloan Fellowship, which I won. Howard Wright was also in Boston on company sponsorship attending an advanced management program at the Harvard Business School.

The radio spewed out a stream of ominous phrases: "Apparently an explosion . . . difficulty maintaining control . . . fast running out of oxygen, water, and electric power . . . Mission Control assures us that the crew is, for now, safe, and has several options for survival." As the announcers stumbled over unfamiliar technical terms and space jargon, there was no mistaking the excitement and concern in their voices. Could this be the night when America's vaunted manned space program would go down to defeat and disgrace, after such a long string of stunning successes, including two manned lunar landings and explorations? The grim prospect loomed of three brave men gasping and suffocating in space while the whole world watched and listened, of their shriveled mummified corpses remaining permanently in orbit as a monument to mankind's overreaching and America's technological arrogance.

Two of the three men who were exposed to the unrelenting peril of space were my friends and professional associates. Fred Haise and Jim Lovell had each spent many days at our Spacecraft Assembly and Test facility in Bethpage, putting the lunar module through its paces against test and checkout computers. The third astronaut, Jack Swigert, I had met briefly in Houston, but I knew he was cut from the same competent, no-nonsense test-pilot cloth as his crew mates. I could picture the three of them, jaws jutting, brows furrowed, as they tried to figure how to work their way out of yet another tight spot. They would be carefully checking all instruments on board the spacecraft, looking through the oxygen and electrical power system diagrams and emergency procedures, and discussing their options in calm, matter-of-fact voices. The imminence of danger would not alter their professional habits.

It was exciting to think that the lunar module might become their lifeboat, the key to their rescue. After the first successful lunar landing, Volkswagen, whose VW Beetle was considered an ugly car, ran a full-page ad in the *New York Times* showing the LM with headlines trumpeting, "It may be ugly, but it gets you there." Where LM was concerned, beauty was definitely in the eye of the beholder, and to me she was beautiful.

Joan routed our oldest son David out of bed to stand watch over the rest of our six children while she drove me to the airport. I was still stuffing LM reference data into my briefcase as we left. It was a beautiful clear April night, and when I met Wright on the tarmac we both strained irrationally to see Apollo up there near the bright Moon. He had talked with some of our people in Bethpage and Houston and determined that NASA definitely planned to execute the LM lifeboat mission, and quickly, as life-sustaining supplies on the mothership command module were rapidly seeping from a mortal wound in its oxygen system resulting from the explosion of an oxygen tank in the service module. There seemed to be no reason not to try the lifeboat approach, for although it had never been rehearsed with either the flight or ground crews or written into specific operational procedures, we had studied the rescue possibility early in the LM's design and had provided additional oxygen, water, and power capacity to cover it.[1] However, to go from a preliminary systems design study done six years earlier to real-time execution of a complex and unplanned sequence of space maneuvers by flight and ground crews untrained in its specifics was quite a leap. We would soon find out whether men and machines were up to it.

During the flight to Grumman's headquarters and main factory complex in Bethpage, Long Island, Wright and I sat behind the pilot in the small, dark cabin of the light plane, lit only by the dim reddish glow of the instruments, looking at the night sky with the moonlit earth below. We talked briefly, trying to reassure each other that the lifeboat mission study we barely remembered had been carefully done and held out the promise that rescue was possible. Then we lapsed into silence, each worrying about whether indeed we could collectively pull it off. In the dark isolation of the cabin, I imagined that I was with my astronaut friends in Aquarius, the mission call-name for Apollo 13's LM. How would they feel knowing their survival depended upon the ability of the LM to perform an emergency mission for which it had not been designed or tested? What could we all do, on board and on the ground, to improve the odds of this gamble? Doubtless it would be terrifying. But wouldn't they feel a gradual growth in confidence as the LM continued to supply sustenance minute after minute, hour after hour, with the hiss of the air supply and whir of the fans providing the same reassurance that I derived from the steady baritone drone of the light plane's engine? And they would know that thousands of engineers and technicians on the Apollo program all over the nation would be exercising their ingenuity to help them meet this challenge. By the time we landed at the brightly lit but deserted Grumman airport I was confident that whatever it took, we would find a way to bring our friends safely home.

We were driven across the runway to Plant 5, where Grumman's Apollo Mission Support Center was located. As we walked toward the front entrance, I met several of my engineering colleagues who were just arriving. Turning around on the front steps, I saw a flood tide of Grumman engineers heading toward the building, anxiety evident in their tense, lined faces. Although it was three o'clock in the morning, it looked like the normal day shift start time of 8:00 A.M.; only the bright moonlight jarred that illusion. No one had asked all these people to come into work—they had simply heard about the problem and decided to do whatever they could to help. It was evidence of the dedication of our Grumman team that I will never forget.

Once inside the Mission Support Center, Wright and I met with the shift leader, John Strakosch, who filled us in on the current situation. During the time it took us to fly from Boston the astronauts had succeeded in stabilizing the spacecraft's attitude, after the oxygen stream had ceased venting from the second, punctured tank. All three crew members were in Aquarius, sustained by the LM's consumables. They were gradually learning how to maneuver the combined command/service modules and LM using the LM's reaction control rockets. Because the mass and center of gravity of the combined modules were so different from that of LM alone, the spacecraft did not respond in anything like the normal manner to which the pilots were accustomed from ground simulations. It gyrated disconcertingly when given straightforward commands to roll, pitch, or yaw.

At NASA's Mission Control Center in Houston, flight controllers sat at their consoles, intently watching the instrumentation readouts from Apollo 13 as they flickered in greenish symbols on their video screens. In a small back room across the hall from the MCC, the SPAN Room, and at a nearby office building, a support team of about two dozen of Grumman's top LM engineers was helping NASA find answers to urgent questions such as, What type of Earth return trajectory should be selected? How much time would it take to return? Would the LM's consumables last that long? What techniques should be used to perform the return maneuvers? Questions requiring research or access to prior test data were forwarded from Houston to Bethpage where more people could be deployed to find the answers.

Our Bethpage Center had four consoles with the same video displays of LM instrument readings beamed from space in real time as did the Mission Control Center. We were able to listen to the flight director, flight controllers, and astronauts on the audio network and talk by telephone with our Grumman people on the scene in Houston. When necessary we could call our subcontractors, suppliers, and consultants from all over the country. This was the first Apollo mission for which I had not been in the SPAN Room in Houston, as close to the center of action as spacecraft contractors were allowed. Despite the greater distance, our access to information and ability to participate in problem solving from Bethpage were excellent.

The estimates of time required for Apollo to return varied between two and a half to four days, depending upon the type of trajectory. The flight director announced over the net that a modified "free return" trajectory had been selected, using the Moon's gravity to whip Apollo around the Moon and send it falling toward the much greater gravitational pull of Earth. Once headed toward Earth, an additional rocket firing would speed up the spacecraft, reducing the return trip to about three and a half days. Two types of information were urgently required: procedures and exact data on timing, pointing, and other parameters for performing the trajectory maneuvers and accurate data of the rates at which each of the life-sustaining consumables was being depleted, together with recommendations on how to conserve them to last until splashdown. It was a good thing that all those Grummanites showed up for work in the middle of the night. Organized according to their technical specialties, the two hundred or more engineers in Bethpage were busy digging out data, conferring with experts by phone, running analyses and calculations, and studying the in-flight and prior ground-test performance of dozens of LM systems and components. We found that the LM batteries were being depleted at an alarming rate, and that immediate, drastic action to "power down" the LM must be taken to survive until reentry. The required power down was far more severe than any of us wanted. It forced shutdown of the inertial guidance system and the resultant loss of on-board data on Apollo's position and velocity, leaving the crew freezing in the dark with only the weakest communication link to Earth still active. After double checking our numbers and scouring the possibilities for less drastic measures, we conferred with our Grumman colleagues in Houston and determined we were in agreement. Bolstered with our independent calculations and concurring opinions, they went forth to convince the NASA flight controllers that less-severe remedies would not suffice.

While debates about how soon and how drastically to power down the LM raged on the floor and in the back rooms of the Mission Control Center, the remaining LM power continued to steadily seep away. A hastily prepared simulation in Houston, using the LM mission simulator, convinced NASA of the need for a sweeping LM power down and was used to develop the switch-by-switch procedures to be read up to the crew. With this argument settled, attention turned to the other less immediately critical consumables (water and oxygen), and to the vexing question of how to perform further return trajectory maneuvers without an aligned inertial platform on the spacecraft, if ground radar tracking data showed error buildups that required adjustment.

Within a few hours, activity in the Bethpage Mission Support Center settled into a deceptively comfortable routine. One by one plans were developed to assure that each of the consumables would not be depleted prematurely. We provided a number of alignment and maneuver procedure recommendations for NASA to consider. From an initial feeling of impending doom, the atmosphere in the room had shifted to one of hope and optimism. By midafternoon I found myself drowsing at the console, and I slipped out to get a little sleep on one of the cots in the nurses' office.

I had not dozed very long before I was aroused by someone calling my name softly and shaking my shoulder. It was Don Schlegel, the shift leader, telling of a new problem: the carbon dioxide (CO_2) level was rising at rate that would exhaust the LM's lithium hydroxide canisters in less than a day. We needed to find a way to use the command module's canisters, but they did not fit into the LM's system.

The problem was that with three instead of two astronauts breathing LM's oxygen, CO_2 was building up 50 percent faster than the LM system design allowed. We would run out of lithium hydroxide, the chemical used to absorb CO_2 to keep it from accumulating to toxic levels. Both the CM and the LM carried the lithium hydroxide in replaceable canisters about the size of a large juice can that were normally replaced every twelve hours. But this was one case we designers had not foreseen. The CM and LM canisters were not interchangeable—theirs was square in cross-section, ours was circular. We literally faced the problem of how to fit a square canister into a round receptacle.

In the Mission Support Center we met with our environmental control system and crew systems engineers, including the resident Bethpage representative of Hamilton Standard, supplier of both the CM and LM's ECS and the spacesuits. Discussions with Grumman and NASA in Houston and with Hamilton Standard at their plant in Windsor Locks, Connecticut, concluded that the best place to devise a solution was at Houston, where accurate mockups and some operating equipment of both spacecraft's systems and the spacesuits were available, and astronauts and the most experienced engineers from NASA and the contractors were on-site. Under the leadership of NASA Crew Systems manager Ed Smylie, a NASA-contractor team in Houston was already at work, with the ground rule that any "fix" had to be something the crew aboard Aquarius could replicate with the materials they had at hand.

From Bethpage there was not much we could do but kibitz from a distance and offer encouragement. The solution they devised was simple but ingenious: instead of trying to square the circle, they used the hoses and fans that normally attach the backpack to the spacesuit to force oxygen to flow through the lithium hydroxide canister, via a jury-rigged adapter made of stiff paper from the flight manual and duct tape. It worked effectively to remove CO_2 in laboratory tests both at Houston and Windsor Locks and was approved by the flight director to be transmitted to the flight crew.

We listened as the complex instructions for constructing the fix were relayed to Jack Swigert by CapCom Joe Kerwin. In the darkened cabin the astronauts used their flashlights to see the parts they were building and assembling. After a final cross-check description to Houston of what they had constructed, they switched the oxygen control valve to place the first command module canister on line. About ten minutes later we cheered when we saw on our screens that the CO_2 level had started to drop. Not a moment too soon, as it had reached thirteen millimeters of mercury, perilously close to the toxic boundary of fifteen millimeters and far above the seven-millimeter level at which the canisters were normally replaced.[2]

With that crisis apparently resolved, we reviewed the overall situation with our colleagues in Houston. There remained almost two days until Odyssey, Apollo 13's command module, would separate from Aquarius and reenter the Earth's atmosphere. Our projections of consumable usage showed enough water, oxygen, electric power, and lithium hydroxide to last until reentry, with the slimmest margins on power and water. We shared the concern of the flight director and the astronaut office about the crew's condition. They were very tired and yet seemed unable to sleep, having been in a constant state of tension for more than thirty-six hours. They were cold and shivering in their thin orange flight suits—with the power down the temperature in the LM cabin had dropped to thirty-eight degrees Fahrenheit, and while conserving water they were also dehydrating. We did not know in Bethpage until after the mission that Fred Haise was also very ill with a kidney infection. He was running a temperature of 104 degrees and at times was very groggy and unresponsive. Such personal details of the astronauts' health were always discussed by the CapCom using the guarded channel, which was not accessible over the mission-support network.

An additional concern was steadily growing: it appeared that another rocket firing would be required to adjust the flight trajectory for the proper reentry angle. When approaching the upper reaches of the Earth's atmosphere, the spacecraft must attain a trajectory angle within a very narrow window—5.3 to 7.7 degrees—in order to decelerate properly from its return velocity of twenty-six-thousand miles per hour. Too steep an angle would result in excessive deceleration forces (Gs) that could crush the astronauts' bodies even as they lay fully supported on their couches and burn away Odyssey's protective insulation, exposing it to fiery incineration. Too shallow an angle would cause Odyssey to skip off the top of the Earth's atmosphere like a flat stone skimmed across the water, sending it roaming through the solar system for eternity.

This was the situation we feared most when we recommended the extreme power down of LM to conserve the batteries. With LM's guidance system shut down, there was no on-board reference of her flight attitude with which to perform the trajectory adjustment rocket burn. Aquarius' position and velocity along her trajectory could be determined accurately enough from the ground-based radars of the Deep Space Tracking Network, but attitude, the direction in which the LM rocket engine was pointing when fired, could only be established by the crew, using some visible reference they could sight upon. We needed to give the crew some practical suggestions on what to use as an attitude sighting reference.

Ever since the LM's guidance system had been powered down, Grumman's guidance, navigation, and control experts had been discussing this problem with their counterparts at NASA and the MIT Instrumentation Laboratory. Whenever someone suggested a technique that appeared to hold promise, it was assigned to an available laboratory to determine whether it could provide the required accuracy. Our Flight Control Laboratory in Bethpage, with its flight attitude table of LM GNC gyros, accelerometers, and other inertial guidance components floating on a frictionless air bearing, was used to check out one such suggestion. From this nationwide fraternal endeavor came a practical solution: the crew should visually align their optical sight with the center of the Earth during the rocket burn, a technique that Jim Lovell had verified sixteen months earlier with an in-flight experiment on Apollo 8. Skillfully keeping the Earth centered in the LM's window during the fourteen-second descent engine rocket firing, Lovell and Haise executed a perfect trajectory adjustment.[3] A day and a half later a second smaller correction was performed by burning the LM reaction control jets while using the same sighting technique, to offset a further unexplained flattening of the trajectory reentry angle.

In the Bethpage Mission Support Center we were exhilarated at the success of the trajectory corrections and heartened by our ability to participate in solving a problem that was far outside our normal scope of activity. Reentry was not a lunar module concern—the fragile LM, with no protective heat shield, must be jettisoned before reentry to burn up like a meteoroid flashing across the night sky. In this time of crisis, the whole space program team across America meshed seamlessly together to do the seemingly impossible. It was a proud moment for all of us.

The time for the crew to leave and jettison Aquarius approached, and there were many procedures to be developed and recommended to NASA for closing out the LM and positioning her at a safe distance from Odyssey for reentry. NASA-Houston was preoccupied with developing and simulating the complex checklist for reactivating the CM, so anything we could do on the LM side to simplify things for them was welcome. NASA noted that Odyssey would be lighter than normal for reentry because she was not carrying the two hundred pounds of Moon rocks and containers that had been expected. What could be taken out of Aquarius to increase reentry weight? We suggested that the crew could cut free some of the fire-resistant Nomex cloth netting and webbing used in the LM cabin for equipment stowage and crew restraint.

As the workload for us in Bethpage tapered off, we became more fully immersed in the impending reentry drama. Of concern to us all was the condition of Odyssey. She had been shut down, cold and dark, for more than three days. Upon reactivation, condensation from the crew's breathing could be expected on all cold surfaces. All wires and electrical connectors would be dripping wet—would they short out? Odyssey's heat shield was another major unknown. It had been directly facing and attached to the service module in which the oxygen tank exploded—had it been damaged by debris? If there were a hole or crack in the heat shield, Odyssey would be torched inside with five-thousand-degree flame, its occupants cremated. There was no way of knowing if the heat shield was still intact.

About five hours before reentry the crew jettisoned the crippled service module. As it drifted away from Odyssey they turned the windows toward it and for the first time saw the full extent of the damage. Lovell reported with astonishment: "There's one whole side of that spacecraft missing."

An entire bay was no longer there, and shreds of debris were dangling from the shattered side, giving eyewitness verification of what those numbers on the screens had been telling us.

The crew transferred into Odyssey and powered her up. No pop or sizzle of electrical shorts, a good sign, and her systems appeared to be normal. After stripping all removable items from Aquarius, the crew closed and locked the hatches and jettisoned their faithful lifeboat. Jack Swigert announced, "Houston, LM jettison complete."

"OK, copy that," said CapCom Joe Kerwin. Then he expressed the proud feelings of all Grummanites for their creation when he added for the world to hear, "Farewell, Aquarius, and we thank you."

Even after being on an emotional roller coaster for so long, I was unprepared for the visceral drama of reentry. Upon reentry the spacecraft is surrounded by a white-hot sheath of ionized gases (plasma) as the atmosphere rapidly absorbs energy from it. During this period of high deceleration, radio signals cannot penetrate the plasma, resulting in a communications blackout. For Odyssey the blackout period lasted four minutes, during which the integrity of her heat shield would undergo its crucial test. When the mission clock showed that the blackout period had ended, we held our breaths waiting for a familiar voice to crackle over the net that all was well. Seconds ticked by, and still no hail from Odyssey. Looking at the TV picture live from the South Pacific, we strained with the sailors aboard the carrier *Iwo Jima* to see, among the scattered clouds, a conical spacecraft dangling beneath a cluster of three large orange-and-white parachutes. After more agonizing seconds, we saw the sailors cheer, and the camera zoomed in on a cloud bank. There they were!

At last Jack Swigert's voice came over the net: "OK, Joe."

Thank God, they made it! The Bethpage Mission Support Center became pandemonium, as all the pent-up emotions of the past three and a half days flooded out in cheers, shouts, and backslapping. On the screens we could see the deliriously happy flight controllers in the MCC breaking out the traditional American flags and cigars. Although we were not similarly supplied, our celebration lacked nothing in intensity and ardor. There were smiles and loud laughter, but not a dry eye in the Bethpage Mission Support Center. The United States manned space program was still number one, and the people who worked for a small aerospace company on Long Island, New York, were a vital part of it. After about an hour of mutual congratulations in the room and over the phone, I suddenly felt overpoweringly tired. Threading my way through the still-excited crowd, I sought out the quiet of the nurse's office cot, and gratefully stretched out. I had had no more than five or six hours sleep in the last three and a half days.

A few weeks later the Apollo 13 astronauts visited Grumman to personally thank the people who built Aquarius and give them keepsakes they made from the materials removed from the spacecraft. Fred Haise had spent much time at Grumman talking to people in every part of the program; as we toured Plant 5 he knew at least as many people as I did. We all were moved by the astronauts' visit and their open expressions of gratitude.

18

The Undaunted Warrior Triumphs

Apollo 14

Alan Shepard was America's first man in space, an honor he had earned by demonstrating leadership in fierce competition with the other hand-picked overachievers who made up Project Mercury's original seven. A virtuoso test pilot, he was intelligent, decisive, and resourceful. And he coveted the chance to be strapped atop a Redstone booster at a time when most large American rockets were exploding shortly after ignition. Such raw courage depended upon a powerful ego and positive self-image, characteristics that also made him somewhat testy. Some associates described him in work situations as arrogant, impatient, and remote, prone to turn unexpectedly on a colleague.

After his pioneering, fifteen-minute suborbital spaceflight on 5 May 1961, Shepard became the first celebrity astronaut. He was honored in the White House Rose Garden by President John F. Kennedy and rode triumphantly down Pennsylvania Avenue to wild cheering by thousands. Even Vice President Lyndon Johnson, who accompanied him in the open car, was amazed at the size and enthusiasm of the crowd.

"You're a famous man, Shepard," the vice president told him.

Shepard's successful flight, and the American public's immense reaction to it, took place as President Kennedy was weighing the decision whether to attempt a Moon program. It appeared to push him over the edge. Less than three weeks after the Rose Garden ceremony, Kennedy announced his decision to make the Apollo lunar landing program a top national priority in his address to a joint session of Congress on 25 May 1961.[1]

I personally saw only brief glimpses of Shepard's complex character. At the M-1 mockup in September 1963, he seemed condescending and amused by the rookie Grumman team and our crude wooden mockup. A year later at the M-5 mockup review, I chanced to share a lunch table with Shepard, Grumman's chief test pilot Corwin H. "Corky" Meyer, and veteran Grumman test pilot Ralph "Dixie" Donnell. The three pilots were in high spirits, witty and charming as they vied with one another in telling about their flying experiences. Shepard told an exciting story about a carrier landing on a stormy night with a rough engine and failing electrical system, when he resisted the impulse to eject and stuck with it to a safe landing. Not to be outdone, Meyer and Donnell told this tale: Corky had been performing gun-firing tests and target practice with the new swept-wing F9F-6 Cougar. The flight test engineers were puzzled by a pressure buildup inside the airplane's nose that they noted during the high-speed gun-firing runs. They were not sure whether it resulted from a leak of external aerodynamic pressure into the nose, or from inadequate ventilation of the gun gas that the four 50-caliber machine guns expelled inside. When he learned that the gun gas was probably flammable, Corky proposed a "quick and dirty" test: installing a few spark plugs inside the nose. If the source were gun gas, it should ignite and produce a pressure spike that would show up on the instrumentation.

The Cougar was outfitted overnight with spark plugs and an ignition system, and Corky took her up to perform the flight test, with Dixie flying a production F9F-5 Panther as chase airplane. When they reached the test altitude in the designated firing range over the Atlantic Ocean, Corky turned on the instrumentation recorders and the spark ignition system and pushed the throttle forward for the high-speed run. As he neared maximum velocity, he fired a long burst from the machine guns. He heard an explosion, smoke filled the cockpit, and the airplane bucked like a bronco. When the smoke cleared, he could see that the aerodynamic nose fairing was gone, leaving the beams and struts of its supporting structure exposed. In the chase plane, Dixie was incredulous, then burst out laughing over the radio, "That was a great test you cooked up, Corky—you just blew the nose clear off the dang airplane!"

Corky hung onto the stick and chopped back the throttle. Despite buffeting and erratic pitch-up tendencies, he was able to land at Grumman's Calverton flight test center. Corky said the test was "a tad inelegant" but certainly answered the question of where the elevated nose internal pressure came from.

Shepard hooted with laughter. "You were absolutely right on that, Corky." Then, grinning broadly and throwing a good-natured wink in my direction, he added, "If you'd left it to the engineers, they'd probably still be trying to work out instrumentation to do the job."

At the top of the growing astronaut corps, Shepard was selected to command the first Gemini mission, with Tom Stafford as his copilot. Then, in the summer of 1963, Shepard was stricken by recurring attacks of dizziness and nausea, which became so severe that he reluctantly reported them to the NASA doctors. They diagnosed his illness as Ménière's syndrome, an inner-ear disorder characterized by excess fluid buildup in the semicircular canals. There was no known cure, although the symptoms disappeared spontaneously in about 25 percent of the cases. Shepard came off flight status while NASA waited to see if he would be one of the lucky 25 percent.

Grounded, Shepard became chief of the Astronaut Office, reporting to Deke Slayton, the other grounded member of the original seven, who headed the Flight Crew Operations Directorate and had his hands full with newly selected astronauts, accelerating Gemini flight schedules, and increasing Apollo activities.[2] Shepard ruled the fractious, competitive pilots with a firm hand, earning their respect if not their affection. He never considered resigning, always hoping that his problem would spontaneously disappear. Finding that he could fulfil his desk-bound duties and still have free time, Shepard became increasingly involved in business, using the many contacts his Mercury celebrity had gained for him.

For six years his situation continued, as the Gemini program was completed and Apollo entered early flight testing. Shepard's business ventures gained him a small empire in shopping centers, hotels, and other enterprises and made him a millionaire. (There was some criticism that he traded profitably on his government-funded Mercury celebrity.) But this was just something to do until he could fly again in space. By the spring of 1968 his condition had worsened, and time was running out to secure an Apollo flight berth. When he learned of a newly developed, risky, and delicate operation that implanted a small silicon tube in the ear to drain excess fluid away to the spinal column, he decided to take the chance. It was his last resort, and it worked.

In spring 1969, with manned Apollo flights occurring every two months, NASA cleared Al Shepard to fly airplanes and spacecraft. Slayton promptly chose him to command Apollo 13, provoking protests from other astronauts that Shepard was being allowed to jump over Slayton's carefully established system of flight crew assignment rotation.[3] Slayton replied that Shepard had always been at the top of the rotation but was on hold while he was grounded.

Shepard's luck had changed more than he knew at the time: his selection for Apollo 13 was overridden at NASA Headquarters by George Mueller, associate administrator for Manned Spaceflight, who thought he needed more training time. Slayton swapped Shepard's mission with Jim Lovell's Apollo 14 crew. Ironically, Shepard's prolonged grounding may have saved his life. Otherwise he probably would have been selected to command the first Apollo flight and would have been inside Apollo 1 on Launch Pad 39.

At this stage of the Apollo program, much more than piloting skills were required of the astronauts. They also had to be competent field geologists, capable of representing the Earth-bound scientists by conducting efficient, perceptive surface traverses in limited time in a hostile environment. Although temperamentally the opposite of the patient, painstaking lunar scientists, Shepard resolved to attain the knowledge and skills needed to be a discerning field geologist. Leading his crew by example, he plunged into field training in the mountains of California and Arizona under Lee Silver, who together with geologist-astronaut Harrison "Jack" Schmitt was giving the Apollo astronauts a crash course in field geology. They learned to identify minerals and to describe what they saw in precise geological terms. They also studied the different theories on the origins of the Moon and how the specific objectives and possible findings of their mission might confirm or refute them. As in all else he did, Al Shepard was resolved to excel in this new dimension of an astronaut's job.[4]

Shepard hand-picked his crew, drawing upon the insight gained into their capabilities from his role as chief astronaut. He selected two rookies who had neither flown in space nor served on a backup crew. He delegated to them the responsibility of becoming experts in the systems aboard their spacecraft, leaving himself free to concentrate on flying and mission operations.

No longer the stern "Big Al" who had lorded it over the others as chief astronaut, Shepard the mission commander was down in the trenches with his crew, soaking up simulator time, conducting ground tests on their command module and lunar module, and clambering up cliffs in geology encampments. His crew mates found him pleasant to work with, even considerate, but still thought of him as a professional associate rather than a pal.

Edgar Mitchell, LM pilot, had a doctorate in aeronautical engineering from MIT and had been an instructor at the Air Force Test Pilot School at Edwards Air Force Base. He was thoughtful and intellectual, soft-spoken, but could become impatient and exhibit a hard edge of anger. He became expert on the LM's systems and their possible failure modes, and I often saw him in Bethpage, roaming the floor in Spacecraft Assembly and Test, examining flight hardware and discussing its idiosyncracies with the technicians who installed and operated it. Mitchell conducted many tests at Bethpage in LM-8's cockpit and was quick to call to my attention any deficiencies he encountered, whether in equipment performance, test procedures, or the Grumman test personnel. His complaints were constructive, involving specific, real problems that we could correct.

CM pilot Stuart Roosa had been an air force pilot in Germany preparing for nuclear war, and later a test pilot at Edwards. Eager and competent, he learned the Saturn booster's systems before his assignment to Shepard's crew, working with von Braun's engineers at Marshall Space Flight Center in Huntsville, Alabama. As CM pilot, he delved into the command and service modules' systems with even greater intensity. He appreciated the free rein that Commander Shepard gave him to spend time at North American's factory in Downey, California, learning the intricacies of his spacecraft.

Shepard's team trained together for nineteen months, longer than any other Apollo crew, due to their reassignment to Apollo 14 and the program delays following the aborted Apollo 13 mission. They enjoyed an easy and correct professional relationship but were not the close buddies that the members of Conrad's and Lovell's flight crews had been. They confirmed Deke Slayton's theory that he could assign any three astronauts together to form an Apollo flight crew, no matter how diverse their personalities, because they were all superb professional pilots.

On 31 January 1971 Al Shepard and his crew, LM pilot Edgar Mitchell and CM pilot Stuart Roosa, were strapped inside the command module Yankee Clipper ready to blast off to the Moon. At forty-seven Shepard was determined to prove that he was the world's best test pilot and astronaut. His Apollo 14 was almost fully dedicated to science and exploration, and Shepard and his crew were committed to every one of its long list of objectives.[5]

After being successfully boosted into Earth orbit by the Saturn 5, and injected into the translunar trajectory by the Saturn's S-4B stage, Apollo 14 encountered a problem that could have ended the mission before it had fully begun. Kitty Hawk was unable to attain hard docking with LM Antares, despite four attempts over a period of one and a half hours.

From my post in the Spacecraft Analysis Room in Mission Control, I discussed the problem with our mechanical systems section leaders, Jiggs Sturiale and Marcy Romanelli. They examined inspection records and photographs of LM-8's docking mechanism and docking ring structure, and had many discussions with their NASA and North American counterparts. Jiggs and Marcy considered debris in the mechanism, or adverse dimensional tolerance buildups the most likely causes of the problem.

Some way had to be found to apply more force to the docking assembly to compress it slightly, allowing the latches on the CM's docking ring to snap home, or the mission would have to be aborted. A solution was required in a few hours, before pressure buildup in the S-4B's tanks would cause them to vent, sending the S-4B, and the attached LM, spinning out of control.

The docking system engineers came up with a modified procedure for Shepard to try. Normal docking procedure was to engage the probe, mounted in the CM's docking tunnel, with the drogue inside the LM's tunnel. Spring-loaded latches on the probe locked into slots inside the mating hole in the drogue, holding the CM and LM together in a "soft-docked" condition. Then an electric motor drive retracted the probe several inches until the spacecrafts' two thirty-two-inch-diameter docking rings made firm contact, automatically activating the CM's twelve docking ring latches to clamp tightly against machined pads on the back side of the LM's docking ring. This produced a "hard dock" in which sizable mechanical forces and moments could be transmitted across the joint between the CM and the LM, allowing the two spacecraft to be maneuvered together under control of the thrusters in the CSM. Kitty Hawk and Antares had achieved soft dock, but they were unable to obtain hard dock.

The engineers' solution was to fire the CSM's thrusters to force the two docking rings tightly together, and then retract the probe while the thrusters continued to fire. The additional compression force applied by the thrusters might overcome the dimensional problem that was keeping the latches from snapping home on the pads.

Roosa lined up the docking target carefully, then fired the thrusters while separated about two feet from Antares. The crew felt a firm thump as the docking rings mated. Shepard retracted the probe. At first nothing happened, but then his panel display flipped to the striped "barber pole" position indicating latches engaged, and the crew heard the reassuring "ripple bang" sound of twelve latches snapping shut. Apollo 14 was still "go" for the Moon.

After the mission, Al Shepard admitted that he had considered riskier solutions if the engineers' fix had not worked. He thought about donning spacesuits, depressurizing the cabin, and removing the probe and drogue from the tunnels for inspection and cleaning. Or, alternatively, using their gloved hands to apply added clamping force to the two docking rings. He was not about to be denied walking on the Moon by such a silly hangup.[6] Recalling his delight at Corky Meyer's quick and dirty test of the overpressure in the Cougar's nose, I found his fall-back docking fixes right in character.

The head-to-head docked spacecraft proceeded smoothly on the long coast to the Moon and were injected into lunar orbit by the large service propulsion rocket engine. Shepard and Mitchell entered Antares in their spacesuits and activated her systems. In the SPAN Room I followed the detailed LM flight plan closely, noting the mission time at which planned actions occurred, such as transfer to LM power, S-band steerable antenna activation, landing-gear deployment, and reaction control system pressurization and hot-firing checkout. As they went behind the Moon and out of touch with Houston, the crew had reinstalled and verified the docking probe and drogue in the tunnels and closed the hatches on the two spacecraft.

A few minutes after emerging from behind the Moon, Antares undocked from Kitty Hawk and revolved gracefully for Roosa's inspection. They moved apart as Shepard conducted a test firing of the LM descent engine and pitched over to observe and photograph the landing site.

I was happily checking off LM flight plan events within a few minutes of schedule when I saw something that made my stomach knot up. The LM guidance computer reported receiving a signal to abort the mission. It could only have come from the abort-stage switch on the LM's control panel. The computer ignored the abort signal because it was programmed only to respond to such a command during powered descent. MIT confirmed that the LGC did receive the errant signal; however, it did not show up on the LM's instrumentation readouts.

I conferred in SPAN with Bob Carbee and Arnold Whitaker, and we alerted our guidance experts and our cognizant engineer on the abort-stage switch. After consulting also with NASA and MIT, we recommended a special procedure to attempt to clear the bogus signal before Antares went behind the Moon: Shepard pressed and held the engine-stop switch while pushing and resetting the abort-stage switch. The errant bit in the computer disappeared just before we lost radio contact.

A whirlwind of activity struck Houston, Bethpage, and Cambridge as NASA Mission Control, Grumman, and MIT debated what to do when Antares emerged again in fifty minutes. In SPAN, Carbee, Whitaker, and I pored over a schematic diagram of the abort-stage switch; in Bethpage two such switches were pulled from the stockroom and installed into test fixtures. At MIT, programmers studied the LGC's software instructions related to abort. The abort-stage switch was a push button with fifteen electrical contacts in a hermetically sealed case. It was a "panic button" for quickly performing an abort-stage maneuver during LM-powered descent—shutting down the descent engine, separating the ascent and descent stages, starting the ascent engine, and pitching up to an ascending trajectory for rendezvous with the command module. Only two signals from the switch went to the computer: one told the computer to shut down the descent engine and start the ascent engine, the other changed the weight and mass properties in the computer to that of the ascent stage only. These signals were of great concern, because if erroneously present during powered descent, the LGC would act upon them. The other thirteen signals activated the explosive devices for stage separation, but they could be blocked by leaving the explosive devices Master Arm switch in the "off" position.

Mission control decided to ask the crew to tap the instrument panel near the switch. After radio contact was reestablished, Ed Mitchell began tapping. Our worst fears were confirmed: as we watched the telemetry readout of the fifteen switch signals, one or another of them randomly changed state (closed or opened) as Mitchell tapped the panel. A solder ball or other conductive debris was floating around inside the abort- stage switch, completing the circuit one signal at a time, depending upon where it alighted.[7] The only solution was to devise a software change that would tell the LGC to ignore signals from the abort-stage switch. MIT had less than ninety minutes to come up with the fix.

Within a few minutes a young programmer at MIT named Don Eyles wrote the corrective software and tested it on the LGC in MIT's laboratory. He transmitted it to NASA and Grumman, where it was independently verified on the LM mission simulator at Houston and the full mission engineering simulator at Bethpage. When Antares reappeared, Mission Control relayed the instructions to Mitchell, who keyed it into the LGC a few minutes before the start of powered descent. Once again the ground-support team had saved the Apollo 14 mission.

Shepard fired the descent engine and began the eleven-and-a-half-minute powered descent. The engine started and burned smoothly, throttling up and down as directed by computer commands, and Antares' altitude above the surface rapidly decreased. About six minutes into the descent the LM passed through thirty thousand feet, and the crew looked for landing radar data. It did not appear, even at twenty-five thousand feet, the specification requirement. The landing radar had checked out okay in lunar orbit. What could be the problem?

Mission rules required an abort if the LM had no landing radar data below ten thousand feet, in about two minutes. I took a deep breath. The LM guidance flight controller recommended opening and closing the landing radar circuit breaker to reactivate the radar's start-up sequence. Mitchell tried it, and good landing radar data filled Antares' display and Mission Control's screens at twenty thousand feet. Carbee, Whitaker, and I cheered and grinned with relief, wiping imaginary sweat from our brows. Another mission save with ground assist!

Shepard pitched over at eight thousand feet and saw his main landmark, Cone crater, dead ahead where it should be.

"Fat as a goose!" he told Houston, and steered Antares to a smooth touchdown, closer to the target than any other landing on the Apollo program. It had been a long, uncertain journey, but the undaunted warrior and his crew mate had arrived on the Moon.[8]

Apollo 14 carried an expanded version of the Apollo lunar surface experiments package and a hand-drawn two-wheeled tool carrier that Shepard called the "lunar rickshaw" (jargon-loving NASA named it the modular equipment transporter, or MET). In the first moonwalk, the explorers deployed the ALSEP and gathered soil and rock samples from the Fra Mauro plain near Antares. They gazed in wonder at the broken, boulder-strewn gray plain, the black sky, myriad bright stars, and stunning blue-and-white Earth overhead. They looked up at the forbidding steep ashy slope leading to Cone crater's rim, which would be the main goal of their second traverse. They selected, described, and documented the rocks and other geological features of most interest to the scientists, living up to their promise to excel at all aspects of their mission.

In SPAN I followed them step-by-step, noting in my copy of the flight plan the actual time at which they performed each lunar surface activity. I heard them report on the Fra Mauro soil, "The surface layer is very soft, fine, and clings to everything. It looks like brown talcum powder."

"The rickshaw is leaving tracks about three-quarters of an inch deep. They glint in the sunlight, and we can see them all the way back to the LM," they said. The astronauts set out an acoustic geophone array and tested it with explosive soil-thumper grenades, leaving a mortar to be fired remotely after they departed. Their precise landing enabled Apollo 14's crew to perform the traverse route and activities they had discussed with the scientists in Houston weeks before.

Once inside Antares, Shepard and Mitchell doffed their helmets, gloves, and life-support backpacks, weighed and stowed their samples, ate dinner, and settled down to sleep. Although ten hours of rest was allocated to prepare them for the physically demanding second moonwalk, eight hours was enough for this pair, and they emerged from Antares two hours ahead of time. They had a difficult climb up the steep slopes of Cone crater, sliding on loose rock and ash, with the MET often tipping over downslope. At times they had to carry it between them. More than 1,100 feet in diameter, Cone crater loomed 250 feet above the plain but was not visible from below. Its rim was expected to be covered with rocks and boulders ejected from deep beneath the surface in the earliest days of the Moon's existence.

Despite a valiant effort, Shepard and Mitchell did not reach the rim and gaze across the crater's vastness. The lack of size reference and memorable landmarks, the sharply undulating surface and unfamiliar lighting conditions combined to make the ascent of Cone's flanks seem like an outdoor hall of mirrors to the frustrated explorers. Mitchell went over to Shepard and showed him the map. "Look," he said, breathing heavily, "let me show you something. . . . We're down *here.* We've got to go *there.*"

They reached a level area, but the rim was still nowhere in sight. Mitchell again studied the map. "This big boulder, Al, that stands out bigger than anything else—we oughta be able to *see* it."

"Okay, Ed and Al." From Houston the firm voice of CapCom Fred Haise told them that time was up. They had used the half-hour extension that Mission Control had granted and still not found Cone Crater. In their stiff pressurized spacesuits, they were tired from the exertion of climbing uphill. It was later determined that during their return they were within sixty-five feet of Cone's rim, which looked to them like just another of many false summits.

Despite not finding Cone's rim, Apollo 14 was a major scientific success. The explorers broke all prior records for time on the Moon (one day, nine and a half hours), duration of moonwalks (nine hours, twenty minutes), and pounds of samples returned (94.4). They squeezed the most they could from Antares, the last of the basic LM designs to fly in Apollo. Subsequent missions would have the extended stay version of LM, capable of spending three days on the Moon, and carrying the Lunar Roving Vehicle.

Before reentering Antares, Shepard produced a golf club and balls, and standing in front of LM's TV camera, he showed the world how far one could hit a golf ball on the Moon, in one-sixth gravity and without atmospheric drag.

"In my left hand," he announced, "I have a little white pellet that's familiar to millions of Americans . . ." On his third try he hit the ball solidly, lofting it up into the black sky and over the craters in slow motion.

"Miles and miles and miles!" Shepard chortled.[9] Without that clumsy spacesuit he felt he could have put it into orbit.

While the crew was tidying up Antares' cabin and eating, in SPAN I was working on a problem in the S-band steerable antenna, the LM's primary communications link to Earth and the only means of high-bit-rate data transmission. The antenna was jittering around its locked-on position when transmissions were sent or received. We feared that it might break lock during ascent, depriving Mission Control of real-time data. NASA and Grumman engineers stopped the jitter by switching to an alternate ground uplink mode and recycling the antenna mode switch in the "auto" position. In any event, the omnidirectional antennas on LM would provide a voice link and low-bit-rate data.

Shepard and Mitchell proceeded through the LM prelaunch checkout. When they test-fired the reaction control system thrusters the exhaust breeze knocked over the erectable antenna that had been used to provide TV and high-bit-rate data and voice during the moonwalks. Liftoff took place on schedule, and the seven-minute ascent engine burn was flawless, placing Antares in the planned ascending rendezvous orbit.

Apollo 14 was the first mission to use a single orbit rendezvous technique, which completed rendezvous in two hours, instead of the two orbits used previously. The new technique worked perfectly and became the standard for subsequent missions. Near the end of rendezvous the LM's abort guidance system, which was shadowing the primary guidance system as a backup, dropped out. Attempts to reset AGS by recycling switches and breakers in the cabin did not work, nor could Mitchell access AGS through its keyboard. Also, the LM master alarm did not announce the AGS failure as it should have. No effect on the mission, but another significant anomaly for our engineers to resolve before the next Apollo launch.[10]

Docking was perfect this time, using the normal procedure, and Shepard and Mitchell rejoined Roosa in Kitty Hawk, bringing their precious lunar samples with them. Antares' ascent stage was jettisoned, and its engine was fired to cause it to hit the Moon midway between the Apollo 12 and 14 landing sites. The blast set off seismic ringing within the Moon that persisted for several hours on both sites' ALSEP instruments. The return mission was routine, and in the Bethpage Mission Support Room I rejoiced in the happy smiles of the astronauts as they strode onto the carrier deck.

On Apollo 14, we in the back rooms had again helped save the mission. NASA Mission Control and its supporting contractors, put to the ultimate test in the incredible Apollo 13 rescue, were a practiced team that responded to problems with aplomb. The resourcefulness and adaptability of a combined man-machine operation in space and on the ground had been demonstrated many times over.

Although the total LM flight anomaly count on Apollo 14 was the lowest yet, the abort-stage switch failure was a serious defect that should have been detected before installation. The initial docking problem showed the need for tighter control and verification of CM-LM interfaces. LM-8 reversed the favorable progression of LM quality and alerted us to improve quality control and take management actions to dispel complacency.

On the Moon's surface, Shepard and Mitchell reached the limits of unaided "walking around" exploration. The Apollo program was ready for a higher level of capability, which the extended-duration missions would provide. Irrepressible Al Shepard and sober-sided Ed Mitchell had been a well-matched lunar exploration team. Now, what would their successors do with a sports car and two extra days to cruise the surface?

19

Great Explorations

Apollos 15, 16, and 17

Thanks to its foresight in upgrading the capability of the later Apollo missions, NASA was positioned to harvest a vast amount of scientific knowledge about the Moon and its origins. The extended-duration missions took advantage of increases in Saturn payload capability, eked out in incremental changes by Wernher von Braun's engineers, and the weight savings we had swiped and scraped from the LM. Lunar stay time was increased to three days, more moonwalks were permitted, the scientific equipment package was expanded, and the lunar roving vehicle was added. The rover promised a new dimension in lunar exploration, as it greatly increased the astronauts' mobility and endurance on the surface. Although the general public was becoming blasé about men on the Moon, scientists the world over anticipated the fruits of great explorations.

The Mountains of the Moon: Apollo 15

In June 1970 I returned to Bethpage from a year as a Sloan Fellow at MIT and plunged into the competition for the newly announced space shuttle, NASA's major post-Apollo program. For the first time in eight years I was not assigned to the LM program, although in spirit I never left it and kept in touch as closely as my shuttle activities permitted. Having directed the engineering design of the extended duration LMs before leaving for MIT in mid-1969, I wanted to see how they had worked out. In August 1970 I attended the design review of LM-10, first of the extended-duration LMs designated for Apollo 15, where the new features were inspected and demonstrated. The redesigned descent stage had larger propellant, water, and oxygen tanks, two additional batteries, an expanded bay for scientific equipment, and a new quadrant bay to house the rover. The ascent-stage changes were minor: added lithium hydroxide canisters, expanded stowage areas for lunar sample containers, food storage, and crew equipment, and accommodations for the redesigned spacesuits.

A demonstration with the completed LM-10 descent stage was held on the Spacecraft Assembly and Test final assembly floor, with astronauts Dave Scott and Jim Irwin participating. All the equipment that the crew would use in exploration was available, either as flight hardware or engineering prototypes. The demonstration was orchestrated by Grumman's Will Bischoff, John Strakosch, and John Rigsby and their NASA engineering counterparts. Working with the engineers who designed the expanded Apollo lunar surface experiment package and the other scientific equipment and tools, they carefully stowed each item into custom designed holders within the bays, making minor adjustments where necessary for a perfect fit. After checking the latching and deployment mechanisms, the bay doors were closed as they would be at Kennedy Space Center prior to launch.

A team of Boeing engineers and technicians showed us the features of their prototype LRV. I marveled when they folded it up like a collapsible stage prop and nestled it securely within the bay. Because the rover was six times heavier than it would be on the Moon, our engineers improvised a counterweight system to ease the jolt when the bay door was opened.

When the descent stage was all buttoned up, Scott and Irwin opened each bay following the procedures they planned to use on the Moon. Surrounded by equipment engineers, they asked questions and gave their comments on each step of the deployment. Most spectacular was opening the rover's bay: Scott and Irwin pulled two lanyards on either side of the bay, the door swung open, and the rover unfolded itself like an insect emerging from chrysalis, ending up angled downward in the bay with its wheels locked in position, ready to be rolled onto the surface. We applauded in appreciation of Boeing's ingenious design.

I returned the following day to watch the final part of the demonstration. Overnight all the chits were dispositioned, final fit adjustments were made, and the descent stage was again closed up as for launch. Scott and Irwin would perform another deployment, but this time in pressurized spacesuits. A technician followed each of them, wheeling a portable cart supplying the suit with air and cooling. This was fascinating; even though they were on the familiar white tiles of the assembly floor, followed by a knot of people in white smocks, the sight of two spacesuited astronauts next to the gold and black foil clad descent stage stirred my imagination. As they bent over and deployed the ALSEP and the rover, I could visualize how they would look doing these same actions on the Moon, stirring up gray lunar dust with every step, surrounded by strange treeless plains and mountains surreally lit by dazzling sunlight or totally hidden in black shadows. This was as close as I would ever get to exploring the Moon, and I relished the fantasy.

I also attended LM-10's customer acceptance readiness review board meeting just before its delivery from Bethpage. NASA's George Low presided, and the board's review took only half a day, with relatively few problems. What an improvement from the early LMs! The review was held in the Plant 25 conference center, which was attractive, comfortable, and provided good acoustics and visuals. About two hundred NASA and Grumman engineers participated in three days of reviews and inspections. Chits were written and dispositioned with the usual rigor, but with less anxiety than I recalled from the early days. The LM program had grown and matured most satisfactorily.

I talked at length with Dave Scott and Jim Irwin. They were very pleased with the quality of LM-10 and excited about their mission. Scott told me the chosen landing site was beautiful and challenging. He hoped the TV images would convey some of its grandeur, because he knew their descriptions would be inadequate, even though they intended to share their sights and feelings with Earth as much as they could during the mission.

The extended-duration LMs greatly increased the scope and efficiency of lunar surface exploration. Only two years after the first manned landing, which at the time had seemed like the ultimate achievement, we engineers had produced a design that opened new vistas of scientific discovery. Three days on the Moon allowed at least three moonwalks, and the rover loosened the bonds of physical exhaustion, enabling the explorers to be out on the surface for up to seven hours at a time, tackling distances and slopes that walking astronauts could not. The rover also saved the crew's strength by carrying their tools and samples and provided precision surface navigation that minimized wasted time seeking landmarks. An advanced TV camera on rover let the whole world share in the thrill of the moonwalks and allowed the lunar geologists in the back room at Houston to take part in their students' field trip and assist with suggestions and evaluations. Lee Silver and the other senior geologists who had tutored the astronauts not only watched them perform but actively participated in the expedition, looking over the explorers' shoulders though a quarter million miles away. The world had never seen anything like it.

Apollo 15's landing site was the most challenging yet attempted by the program, and a very beautiful and dramatic area. It was situated on a level plain bordering Hadley Ridge of the lunar Apennine Mountains, between towering eleven-thousand-foot Mount Hadley Delta and Hadley Rille, a sinuous three-thousand-foot canyon. Scott and Irwin were startled when the LM Falcon pitched over at nine thousand feet to see the brilliant sunlit flank of Mount Hadley Delta above them to the left. The scene ahead was unfamiliar, and Mission Control informed them that the guidance system had put Falcon three thousand feet south of where it should be. Thanks to the prominence of Hadley Rille and Mount Hadley Delta, Scott was able to correct their flight path manually and land very close to the target point. Dust obscured his vision for the final sixty feet, making this the second LM to touch down on instruments. In view of their intended three-day stay on the Moon and the long duration of their planned moonwalks, Scott and Irwin planned to sleep before setting foot on the Moon. But first Scott conducted a visual reconnaissance of the area, as Lee Silver had taught him to do at any new field site. Depressurizing Falcon's cabin, he raised himself halfway out of the upper hatch, and for half an hour he gazed at and photographed scenes of unsurpassed beauty and grandeur: the rounded gray flanks of Mount Hadley, covered like a ski slope with the untouched snows of eons of cosmic dust, and the winding, mysterious depths of Hadley Rille, perhaps holding a key to the Moon's history written on its walls. He was thrilled at the prospect of exploring such a vista.[1]

Of the many engineering improvements introduced on Apollo 15, among those most appreciated by the astronauts was the redesigned spacesuit, which was more mobile and flexible and easier to doff and don, allowing the crew to remove their spacesuits in Falcon's cabin and enjoy the unfettered comfort of their flight suits while eating and sleeping. This did not help my personal uneasiness about micrometeoroid penetration of the LM's cabin while on the Moon, but after the prior missions I was feeling more confident and worried less while they slept.

During the Apollo 15 mission, I dropped into the Bethpage Mission Support Room to watch critical mission events, and after-hours I watched the explorations, which were shown almost continuously from the rover's camera over one of Mission Control's channels. I saw most of the second moonwalk and felt more like a participant in the adventure than ever before. The rover climbed three hundred feet up the flank of Mount Hadley, maintaining good speed but giving Scott and Irwin concern about tipping over on the steeper slopes. At one point while both astronauts were gathering samples on foot, the rover started to slide downhill, but Scott quickly grabbed and held it. Scott panned the rover's camera over the scene from their highest point, showing the rounded hills golden in sunlight, with the darker plain and light-walled Hadley Rille below. The LM Falcon was far away, a tiny speck in the unearthly panorama. (Mission rules limited the crew to driving no farther from the LM as they could safely walk back—about six kilometers.) The mountains of the Moon were hauntingly beautiful and mysterious, and palpably ancient.

On the way back down Mount Hadley, Scott and Irwin explored midsize Spur crater, collecting and documenting samples. An unusual whitish rock caught their eyes; when Scott brushed it off he could clearly see the white crystals of anorthosite, most likely from the Moon's primordial crust. Knowing the value and import of the treasure they had found, Scott and Irwin displayed it before the camera and exulted with their scientist colleagues in the back room at Mission Control. (This sample, dubbed the Genesis Rock by a reporter covering the mission in Houston, was found to be 4.5 billion years old, probably dating from the formation of the Moon.) Returning to Hadley Base, Scott made another major contribution to science by drilling a ten-foot core sample tube into the surface, despite stubborn resistance and the pain of aching fingertips, which had been pressed too long and hard inside his pressurized gloves. (The core sample was worth the effort required to collect it. Scientists identified forty-two layers of soil, the bottom layer undisturbed for half a billion years.)

Their third and final excursion took Scott and Irwin to the edge and some distance down the sloping side of Hadley Rille. They saw and photographed layering of the canyon walls from repeated lava flows, providing convincing evidence of the active volcanism that played a part in shaping the Moon's ancient past. They collected many more rock and soil specimens and retrieved the ten-foot core sample. Before entering Falcon, Scott parked the rover nearby and pointed its camera to capture Falcon's liftoff.

As I saw it from the MSR in Bethpage, the liftoff was amazing. The ascent stage leapt upward very quickly in a shower of silver and gold shards of torn insulation and disappeared from the camera's field of view. For a few seconds bits and pieces fluttered to the ground, and then the LM descent stage and those ALSEP instruments in the picture were still—frozen on the Moon for eternity. I still picture them that way whenever I look up at the Moon. Six silent sentinels awaiting the return of the next wave of lunar explorers.

Ascent and rendezvous were smooth and uneventful, and Scott and Irwin were reunited with Al Worden, who had made his own major contribution to lunar science. The scientific instrument module on Endeavour had mapped and examined much of the Moon's surface and recorded copious data on its composition and characteristics. These explorers were returning with their spacecraft overflowing with astutely selected samples and data that would add greatly to mankind's knowledge of the origins of both the Moon and, by proximity and analogy, Earth. Safe aboard the recovery carrier *Okinawa* in the Pacific, they breathed again the cool fresh air of Earth and delighted in mingling with their fellows, the postmission quarantine regimen having been dispensed with as unnecessary.

I met the Apollo 15 crew several weeks later at a dinner in Houston. Dave Scott and Jim Irwin shared with me many details of how things felt and sounded within the LM. They described liftoff and ascent as a smooth elevator ride, with the moonscape rapidly shrinking in the triangular windows, and the flaming explosive energy of the rocket engine burning only inches behind them as a steady, nonthreatening vibration transmitted mainly through their feet. They mimicked some of the sounds aboard LM: the sharp bang of the cabin depressurization valve, the whines and hums of the ECS pumps and fans, and the abrasive grinding of the steerable communications antenna. They willingly answered all my questions and repeatedly returned to their feelings of gratitude at being able to witness such wondrous beauty in another world. Thanks to their openness and sharing, and the added dimension provided by the rover's TV, I felt deeply involved with the adventure of this unprecedented mission. There was much more to my pleasure in Apollo 15 than satisfaction in our LM's near flawless performance. For the first time, my fantasy of stowing aboard the LM and exploring the Moon with the astronauts had found a degree of fulfillment.

The Central Highlands: Apollo 16

In April 1972 Grumman was at a fever pitch of preparation for the space shuttle competition. NASA's request for proposals was expected within a few weeks, and we had some six hundred engineers busy completing our studies and analyses and developing our main proposal themes. As deputy director of the space shuttle program at Grumman, I bore a major responsibility for the proposal, which meant long hours and weekend work. This left not much chance for me to follow the Apollo 16 mission as it unfolded, although I did try to drop into the Bethpage MSR after-hours to watch the mission control monitors, if only for a few minutes. Some nights I just had to settle for the truncated news summaries of the mission on broadcast TV.

Apollo 16 was targeted at the Descartes Highlands, near the center of the Moon north of the equator. The geologists thought the bright, extensive highland areas might have been created by volcanism in the Moon's early history, predating the lava flows that created the Maria. After carefully studying highland landing sites, NASA concluded that the Descartes region, although very uneven and cratered, offered enough level areas to be safe for a lunar module landing. Scientists were eager for their first exploration of terrain that appeared representative of a major portion of the Moon's surface.[2]

Apollo 16 had an all-southern crew with a "down home country" style: Comdr. John Young, LM pilot Charles "Charlie" Duke, and CM pilot Thomas Kenneth "Ken" Mattingly. They arrived in lunar orbit uneventfully and on schedule, but then an unexpected problem threatened the mission. After CM Casper and LM Orion separated and Orion preened for Mattingly's predescent inspection, Casper failed one of the checkout procedures Mattingly put it through. When he checked the secondary gimbal control system[3] in the service propulsion system, the steering gimbals oscillated instead of holding a steady commanded position. Although the primary gimbal control had checked out, mission rules required that both systems be operational before permitting LM to descend to the Moon. Casper and Orion held their positions in close orbits for more than six hours while Mission Control worked on the problem. They finally gave a mission go-ahead, based on the judgment that even with the positional oscillations of the gimbals, the secondary system could safely control an SPS engine firing.

While in the holding pattern Orion also developed trouble. Houston flight controllers noticed that the pressure in the reaction control system fuel tank was creeping upward out of limits. Grumman's Will Bischoff, Manning Dandridge, and Ozzie Williams decided this was caused by leakage in an RCS pressure regulator and recommended opening the ascent propulsion–RCS interconnect fuel valve. This allowed the excess pressure to relieve into the much larger ascent fuel tank and solved the problem.[4]

Young flew Orion to a smooth landing in rough, heavily cratered terrain, near the base of towering Stone Mountain. He and Duke conducted three highly productive moonwalks, the first two exceeding seven hours duration each, the third exceeding five hours. They drove the rover up Stone Mountain to a point five hundred feet above the plain and gazed with wonder on an otherworldly panorama. Their home base LM was not even in sight, hidden by an undulating fold of Stone Mountain's slope. The mountain's underlying rounded shapes recalled Mount Hadley, but Stone Mountain was more rock-strewn and rugged. They found many whitish crystalline rocks, anorthosite from the Moon's primordial crust, like the Genesis Rock. But contrary to the geologists' predictions, they found no evidence of volcanism in the highlands. The area appeared to have been formed by intensive meteorite bombardment in eons past. The two explorers were excited and expressive, punctuating their comments with "Wow!" and "Lookit that!" Young and Duke freely shared their sights and feelings with the world, in plain-spoken American country English, to the delight of much of their U.S. audience.[5]

The explorers returned more than two hundred pounds of lunar samples, a new record, plus data and photographs from the surface and from the orbiting scientific instrument module. Mattingly performed a spacewalk to retrieve film from the SIM at the base of the service module, in the daunting, spectacular emptiness of space almost halfway between Earth and Moon, an experience forever seared into his memory.

Grand Finale: Apollo 17

Joan and I and our three youngest children, Christopher, age twelve, Jennifer, nine, and Peter, seven, waited for hours in the bleachers of the VIP viewing area through prelaunch holds in the countdown of Apollo 17, the first nighttime launch to the Moon. As the December chill and dampness deepened, we watched the Apollo stack in the distance on Launch Pad 39A, bathed in the glare of floodlights and exhaling streams of white vapor, as though nearly bursting with stored energy. The children grew restless and ran around the parking lot, emptying the small cooler of snacks and finding beetles and other insects in the weeds. Well after midnight, when the countdown passed the planned twenty-second hold, silence fell over the watching throng, and I squeezed Joan's hand for reassurance.

The dazzlingly bright yellow-white plume spilled out from beneath the Saturn booster and, as the black-trimmed white cylinder cleared the tower, expanded until it looked like the rocket was riding astride the Sun. The nighttime darkness vanished in the harsh light of a klieg-light white dawn, showing the flat expanse of dunes, wetlands, and tropical scrub growth to the far horizons. I gaped open mouthed at this manmade wonder, and forgot about the oncoming shock of sound until it hit my ears with staccato fury and sent earth tremors up my legs. Then as we stared in wonder, the Sun slowly transformed into a brilliant star and the shrinking rocket came less distinct in its glare. For several minutes our eyes were riveted skyward, until the sound bombardment faded to a distant rumble and the departing moonship was a dimming star. Darkness reclaimed its rightful reign over southern Florida, but the curving trail of fluffy white exhaust cloud, luminous in the light of the target Moon, attested to the reality of the miracle we had all witnessed.

As we waited on the bus after the launch, stuck in the traffic of a million departing viewers, our kids were so excited by what they had seen that they chattered endlessly, not settling down for sleep until long after we returned to our motel. Before finally dropping off, little Peter asked if the astronauts were sleeping too in their spaceship. For all of us it was a night to remember.

I had come down to Kennedy Space Center a day before Joan and the children to watch the final launch preparations from Grumman's Mission Support Room in the Operations and Checkout building. In the afternoon before launch, Grumman's KSC director, George Skurla, told me this story: NASA's Apollo launch director, the hard-driving Col. Rocco Petrone, had called Skurla into his office a few months earlier and asked what he was doing to maintain morale and discipline among the Grumman team at KSC. With no follow-on space work, the Grumman people were literally working themselves out of their jobs on Apollo 17, which worried NASA management. Would their quality and dedication be sustained at the same high standards of prior launches? Skurla told him, "Don't worry about my Grummies, they'll do just fine."

On launch day Petrone summoned Skurla into his office and showed him a poster left on the platform on the launch tower outside the white room and the command module, where it was hoped the astronauts would see it as they boarded their spaceship. "Don't your people ever learn? This is a violation of the rules," he said. On a previous mission he had demanded that the perpetrators of a similar infraction be fired.[6] On this final Apollo launch, his displeasure was more measured. He looked at the poster with Skurla, which was signed by each member of the Grumman launch team: "This may be our last LM, but it will be our best!" Petrone's face broke into a half-smile and he admitted that he admired their spirit.

Earlier that day I had rendezvoused with Joan and our children at a large prelaunch luncheon sponsored by Grumman. Several astronauts' wives attended, including Jan Evans, wife of Apollo 17's CM pilot, Ron Evans, with whom Joan was acquainted from prior mission visits in Houston. Some NASA officials also attended, including Bob Gilruth and his wife, and Gen. and Mrs. Carrol "Rip" Bolender. Gilruth offered brief remarks praising Grumman for its outstanding contributions to the Apollo program. Jo Evans, wife of Lew Evans, who had died suddenly in June, also said a few heartfelt words thanking Grumman's LM people for keeping Lew's spirit alive. The group faced the upcoming launch confidently, with pride and relief, and regret that this was the finale.

The valley of Taurus Littrow, the place chosen by the scientists as likely to fill in many still-missing pieces in their story of lunar creation, was a complex and challenging site. It was a small box canyon on the northeast edge of the Sea of Serenity, about four miles wide, bordered by two seventy-five-hundred-foot sheer walls of high plateaus, known as the North and South Massifs. Researchers were attracted by its great geological variety: it was heavily cratered and rock-strewn, and the bases of the massifs were piled high with boulders and landslide debris from the walls. Orbital photos suggested that its light-colored surface was dusted with a thin layer of darker material, which experts thought could be volcanic ash or impact ejecta. The landslide debris from the base of the massifs could contain older upthrust material from the Moon's interior. The area was a geological treasure trove that held far more than could be explored in three excursions, even with the rover and with a geologist-astronaut on the scene.

The crew of the final Apollo expedition included LM pilot Harrison H. "Jack" Schmitt, the only trained scientist to fly in space on the Apollo program. With Comdr. Eugene Cernan, a space veteran of the Gemini 9 and Apollo 10 missions, and CM pilot Ronald Evans, they were very capable lunar field geologists as well as astronauts.

Apollo 17 was a record-breaking, extremely productive mission. Cernan skillfully guided the LM Challenger to a pinpoint landing on the pockmarked floor of the narrow valley, enabling the crew to make maximum use of their planned traverse routes. They gazed awestruck at the cornucopia of geological specimens laid out for their inspection. Schmitt's expertise let them wisely prioritize their time in sampling this vast selection. They were immersed in spectacularly beautiful scenes, dominated by the dazzling white walls of the massifs and the darker, rubble-strewn valley floor. An iridescent blue-and-white crescent Earth hung suspended above the towering flank of the South Massif, eliciting an exclamation of astonishment from Cernan.

As befitted the finale, they broke all prior Apollo mission records—for time on the surface (three days, three hours), longest moonwalks (three, each over seven hours), greatest distance from the LM (5.5 miles), and lunar samples returned (243 pounds). While exploring conical Shorty crater they discovered orange soil and took many samples. This was later determined to be composed of glass beads of many colors—orange, green, blue, even brown and black—and of unique chemical compositions, rich in titanium and iron but containing little silica. Scientists attributed them to "fire fountain" eruptions of pressurized, gas-saturated lava from deep within the Moon during the period of active volcanism.[7]

On their third excursion the explorers reached a huge dark boulder on the side of the North Massif—a preselected destination, as it was visible in the Apollo 15 photographs, along with the long trail it had left when it rolled down the massif's wall. It turned out to have broken into five large pieces, and Cernan chipped the micrometeorite patina off its surfaces while Schmitt examined it and discussed what he saw with the scientists in Houston. It was an extremely complex boulder, containing intermingled masses of tan-gray, blue-gray, and other rocks and chips of white crystalline minerals. Cernan and Schmitt examined and photographed it at length, chipping off many samples, because they knew that this boulder had a dramatic story to tell about the Moon's evolution.[8]

Both spacecraft performed perfectly throughout the mission. As the Grumman launch crew had promised, their last was their best. In a brief ceremony before entering Challenger from the surface for the last time, Cernan showed the worldwide TV audience the commemorative plaque on the front landing leg and challenged future generations to follow soon in their footsteps.[9] It was official; the enormously successful program of manned exploration of the Moon was over, not because all the questions were answered, but because time, resources, and will had run out.[10]

Watching parts of the moonwalks from the VIP viewing room in Mission Control at Houston, I savored the marvels of exploring another world by proxy. It felt strange to be a visitor to an Apollo mission, and I frequently slipped down the hall into SPAN, where Coursen, Bischoff, and Carbee presided for Grumman. It was a relaxed atmosphere, with no spacecraft anomalies to be worked, and a feeling of pride and satisfaction spread over the usually anxious engineers.

I attended bittersweet celebrations in Houston with Grumman and NASA colleagues. After so long, it was ending, but what a triumphant way to ring down the curtain! I returned home to see another safe, precise carrier recovery on TV. Then it was undeniable—Apollo was over. Whatever would I do for an encore? I knew there would never be anything like it in my lifetime. And what a ride. Twelve glorious years of purpose, dedication, and achievement. How ironic that it ended as the growing tragedy of the Vietnam War was tearing our country apart and alienating a generation from the ideals of patriotism and national purpose that had made Apollo possible.

Not one to dwell in the past, I took up the daunting challenge of trying to keep Grumman's space business afloat in a shrinking market environment. From Apollo I carried lessons of motivation, teamwork, and quality throughout my remaining career. Although there would never be another Apollo, I resolved to apply what I had learned there to the next generation of aircraft and spacecraft.

20

Our Future Slips Away

For a brief interlude in 1969–70 I left the relentless pressure of the LM program and sampled the treasure house of university knowledge. The year I spent at MIT studying industrial management as a Sloan Fellow was stimulating and informative. I learned many aspects of the theory and practice of management and formed a broad network of friends and acquaintances. My whole family came with me, into a big rambling frame house in Winchester, Massachusetts, that bordered on the town forest—a sylvan enclave only eight miles from the Boston Commons. The change of pace was welcome, for although student life was demanding, it was more flexible than work on the LM program had been, allowing us to spend more time together as a family and explore Boston and New England. I kept closely in touch with Grumman and made frequent visits to support the flight missions. In the spring Joe Gavin and Grant Hedrick told me that when I returned in June, they wanted me to join the rapidly growing space shuttle program.

The space shuttle was NASA's major post-Apollo manned spaceflight program. It had supplanted Wernher von Braun's dream of a space station as NASA's primary objective because planners reasoned that unless the cost of traveling to and from orbit could be reduced, long-duration, manned activities such as a space station would be prohibitively expensive. In management strategy meetings led by the forward-looking George Mueller, NASA opted instead to lower the cost of spaceflight by developing a reusable, round-trip booster system.

The space shuttle program's goal was to provide low-cost access to space. It was envisioned as a fully reusable system, with winged, rocket-powered booster and orbiter stages mounted piggyback at liftoff. The shuttle would be launched vertically like a rocket, but each stage would land horizontally like an airplane. The booster stage would exhaust its rocket propellants and detach from the orbiter, then the orbiter's rocket would propel it into Earth orbit in a manner similar to a multistage expendable booster, except that the orbiter's engines would be fired continuously from liftoff. Both stages would have human pilots and would land using retractable turbojet engines and landing gear. Their wings and reaction control rockets also provided orbital maneuvering capability on reentry, allowing each vehicle to change its return trajectory by hundreds of miles to reach the landing site.[1] The orbiter had a large cargo bay and was required to carry a maximum sixty-five-thousand-pound payload into low Earth orbit. This heavy lift capability further reduced the dollars-per-pound cost of the program. The fully reusable shuttle was a huge, complex, and technically demanding system that would be very expensive to design and develop, even if its recurring operational costs met NASA's ambitious goal of one thousand dollars per pound to orbit.

NASA held a competition in late 1969 for studies to define the space shuttle and its missions. Study contracts for $8 million were awarded to North American and McDonnell Douglas, and the companies augmented this amount liberally from their own discretionary funds. Several hundred engineers worked on each study team, producing many detailed design concepts on their most favored approaches.

Grumman formed a space shuttle team under the direction of veteran project engineer Larry Mead, with me as deputy and Fred Raymes as proposal director. Working with Max Faget and his group, we obtained a $4 million alternate space shuttle concepts contract to explore and develop less complex and costly alternatives to the fully reusable systems. This put us into direct competition with NASA's mainline phase B studies, which were designing fully reusable boosters and orbiters in increasingly fine detail.[2]

The paradigm-shattering concept that made the space shuttle feasible occurred late in 1971. My recollection is that Faget and his team thought of it first then showed it to us, but some of my own engineers may dispute that, since our work was so closely intertwined with NASA's. The concept eliminated the booster completely, removing all main propellants from the orbiter and placing them in a very large, lightweight expendable tank to which were attached two large, reusable solid-propellant rocket motors. This simplified the system and reduced its cost, reduced the liftoff weight, and reused most of the high-value items. The solid rockets were lowered into the ocean by parachute, where their empty steel cases would float until they were towed ashore, cleaned, and refilled with solid propellants for another flight. The concept was a logical evolution of our investigations of external drop tanks on the orbiter, first hydrogen, then also oxygen, and strap-on solid rockets, which kept growing bigger in our designs.

I was elated. If this alternate concept proved viable, it would negate two years of heavily funded studies by our competitors. We were within reach of repeating our successful pre-LM tactic: redefining the mission and the spacecraft configuration into something we knew more about than anyone else. Would the lightning of innovation strike twice?

We redesigned the orbiter for the new concept. We removed all cryogenic propellants (i.e., hydrogen and oxygen) from the orbiter and used high density storable liquid propellants (as on the LM ascent and descent propulsion systems) for the orbiter's orbital maneuvering system (OMS, the equivalent of LM's propulsion and reaction control systems). The external tank was jettisoned shortly before reaching orbital velocity, descending to destruction in fiery reentry and splashdown in the Pacific Ocean, while the two six-thousand-pound-thrust OMS engines carried the orbiter upward into orbit. Our aerodynamic design studies converged on a sharply swept "double delta" wing configuration smoothly blended into the fuselage, with relatively blunt leading edges to hold down peak heating. This design produced a sufficiently high lift/drag ratio to meet the air force cross-range maneuver requirements. Further analysis showed that this design had sufficient post reentry L/D and glide maneuverability to reach the airfields at Edwards Air Force Base or Kennedy Space Center from orbit without turbojet power, making "dead stick" landings with acceptable touchdown velocity and landing approach maneuvers. Turbojet engines were not needed, simplifying the design and eliminating jet fuel from the orbiter.

With these innovations and refinements, the shuttle design converged into an attractive package. The revised orbiter was much smaller and safer than it previously had been, with no low-density, explosive cryogenic propellants stored inside the fuselage. The evolved alternate concept appeared practical to design, build, and operate. It was predicted to meet the cost targets and all the technical requirements except for full reusability. It nominally met the air force's requirements for payload capability, cross-range maneuverability, and launch and landing sites, giving no cause for the air force to retract support of the program. As each new piece of the design puzzle fell into place, I felt the irrepressible excitement that accompanies the discovery of a winner.

On 5 January 1972 the White House announced that NASA would develop a space shuttle based upon the alternate "stage and one half" design. I was euphoric, along with most of Grumman. Our hard-fought strategy had worked, and our competitors had to scramble to master the technical intricacies of the alternate concept. Their prior lavishly detailed studies and designs were largely scrapped. It seemed like a replay of the LM competition, except that this time Grumman had an unsurpassed track record of Apollo LM performance, a large, space-experienced engineering and manufacturing work force, and state-of-the-art space facilities.

The space shuttle request for proposals was released in May 1972, starting a nonstop frenzy of strategizing, writing, and editing. It was a large proposal, restricted to four thousand pages for the technical and management volumes, but with unlimited pages for financial and cost estimating data, and was due in sixty days. The daily proposal team meetings, at which Mead, Raymes, and I presided, grew in attendance and intensity, as enthusiasm mounted and many specific short-range tasks had to be accomplished. Shortly before the RFP was released we moved the shuttle program from the third floor of Plant 25 in Bethpage to a large proposal center in a new leased office building in the Huntington Quadrangle in Melville. There we had spacious quarters on a brightly lit floor with large peripheral windows, recessed fluorescent lighting, tile flooring, office cubicles and furniture, and many conference rooms, all fresh, clean, and attractive.

Without warning, our charismatic president Lew Evans died. He had a massive heart attack at home; it killed him instantly. Although Evans had suffered a heart attack a few months earlier and had been hospitalized briefly, he seemed to have fully recovered, so no one was expecting what happened. Grumman was plunged into unrelieved shock and gloom. Evans was widely liked and respected at the company, and Grummanites at all levels felt a profound sense of loss. Thousands paid their respects at his wake and funeral, with the Grumman security guards helping the Nassau County Police direct traffic. I felt absolutely terrible; I had lost a good friend, and a man whose uplifting, buoyant spirit had made work enjoyable. But for the need to complete our shuttle proposal, I could have lapsed into unproductive depression.

Beyond my personal feeling of loss, I realized that Evans's death was a severe blow to our shuttle proposal. Of all Grumman's executives, Lew had established the closest personal rapport with the leaders of NASA, especially Low, Mueller, Gilruth, and Phillips. They trusted him and believed they could count on Grumman to do the right things while he was in charge. Joe Gavin, who took over for Evans, commanded NASA's trust and respect also, but he was not as involved in political strategizing to promote the space shuttle program.

NASA was doubtless concerned about the loss of Evans, and not just its effect upon Grumman's ability to live up to its shuttle proposal but on its survival as an independent company. This was a result of serious problems on the F14 Tomcat supersonic fighter aircraft program. The F14 was one of the last "total package procurement" (TPP) programs, an ill-conceived Department of Defense management initiative that required contractors to sign up for the total package of a major weapons system development and production program, including production lots many years in the future, under a firm, fixed-price contract. Minor allowances were made for cost inflation, based upon the low (2 percent or less) inflation rates of the 1960s. However, the ink was barely dry on Grumman's F14 contract with the navy when inflation began to surge, from 6 percent in 1969 (fueled by the intensifying war in Vietnam) to more than 22 percent in 1973 with Vietnam plus the Arab oil crisis. With mounting inflation, the DOD soon abandoned TPP for future procurements, but Grumman was stuck with a large fixed-price contract that had the potential to drive it into bankruptcy.

The F14 contract was a featured topic at hearings of a subcommittee of the Senate Armed Services Committee in April 1972 (shortly before release of the shuttle RFP by NASA) that produced a prolonged daily barrage of negative publicity for Grumman. The company was vilified as a greedy, bloated contractor seeking a government bailout to obtain excessive profits. Senator William Proxmire frequently charged Grumman with mismanagement, poor quality, and low performance. Although directed primarily at the F14 program, these allegations produced a broad negative image of Grumman. It was certainly not the environment we needed on the eve of a major government procurement decision. Because 1972 was a presidential election year, the DOD and the navy hung tough, and there was no financial relief in sight for Grumman as NASA approached the space shuttle contractor selection.[3]

Despite these grievous blows of fate, we managed to produce what I considered an outstandingly good proposal. The quality of the writing, graphics, and illustrations was uniformly high. Our primary themes were echoed throughout in many specific variations, backed up with evidence drawn from all technical and management disciplines: Grumman was innovative and reliable and delivered high-quality products, as shown on Apollo, and as we played a primary role in helping NASA define the space shuttle system and missions, we were best prepared to design and build it. We emphasized our systems engineering and program management capabilities, proven on LM and on our major navy aircraft programs, and we proposed a carefully structured management plan for shuttle. Our technical approach was conservative, building upon our successful LM experience. We proposed a thorough plan for engineering analyses and tests, which was technically satisfying and logical but costly. As I flipped through the pages of the finished volumes the day before shipment, I felt that we had given this proposal our best effort and could be proud of the result. I was also exhausted and slipped away for a couple of days, relaxing at home before going back to face the next immediate challenge, our oral briefing to NASA's Source Evaluation Board.

Our shuttle orals were thoroughly rehearsed and well choreographed to retain the listener's interest and engage his intellect as well as deliver our sales themes and messages. Gavin and I gave most of the briefing, using high-quality visuals (thirty-five-millimeter slides) prepared in our proposal center. By the time we left for Houston I had most of my pitch memorized.

I felt confident and comfortable, and my briefing went very well, as did Gavin's. It seemed to me that I had the audience enraptured; their attention never wavered, and you could see their knowing smiles and glances as for point after point I offered solid engineering data supporting my claims. I wound up in a breathless crescendo of earnest entreaty as the clock struck the hour, prompting applause from the normally deadpan board. After we were ushered out, Dave Lang and I danced a jig of joy and relief on the front steps of Building 1, while singing "Hello, Dolly." The warm reception made me and many others think we had a winner, and that our orals briefing had hit a home run.

In our best and final offer (BAFO) to NASA we reduced our estimated cost from $4 billion to $3.2 billion by cutting back on our development test program. After BAFO the atmosphere in Washington was not encouraging. The F14 stalemate continued, with the navy adamant and Proxmire continuing well-publicized attacks on Grumman. I was nervous but hopeful, convinced that we had won the technical contest and would narrowly prevail in the management and financial areas. Basically I believed that our Apollo record was too outstanding to go unrewarded. I recalled a late-night discussion with two of the German engineers from Dornier Systems during the proposal. Finishing our work for the night, we began to philosophize a bit. One of the Germans asked me, "What will Grumman get on shuttle if you lose this competition?"

"Nothing," I replied.

"Surely, with all this talent and expertise, your government would not allow it to go to waste if you lose. What then would you get? What is your loser's portion?"

"Nothing," I said again. "This competition is totally winner takes all."

They both stared at me unbelieving. Then one shook his head slowly and said, "Then America is indeed a very rich country. But can even a rich country afford such waste?"

Shortly before President Nixon's reelection, the space shuttle contract was awarded to North American Aviation for $2.6 billion cost plus incentive fee. I was stunned and dejected, and for a while fell into pointless recriminations and what-ifs: we should have been more aggressive in simplifying our proposed program and cutting its cost; maybe our management structure was too complex; we should have been more persistent in trying to strike a deal with North American Aviation, and so forth.

I never learned how NASA's decision was made. The official debriefing said that North American won because it offered a superior design and lower cost. Persistent rumors said the Source Evaluation Board had rated Grumman highest, but the source selection authority (SSA), NASA Administrator James C. Fletcher, had selected North American Aviation. The official SEB record showed that NAA was rated highest, Grumman a fairly close second, and McDonnell Douglas third. All the competitors were close enough in their SEB scores that the SSA had a free hand to choose the winner.[4] The SSA would have had much explaining to do with Congress and the American public if he had selected Grumman, given its F14-tarnished image. NASA management was doubtless concerned that with Evans gone and the company facing huge financial losses, Grumman would not survive, making its ability to live up to its shuttle proposal questionable. President Nixon was thought to favor North American Aviation because it would create jobs in California, and because Col. Willard Rockwell, the chief executive officer of North American Rockwell, was a longtime, outspoken political supporter and contributor ever since Nixon's early, controversial days in the House of Representatives.

Some in the NASA hierarchy may have remembered the dark side of their relationship with Grumman on Apollo, in which Grumman appeared arrogant, holier-than-thou, a loose cannon. These executives felt that Grumman had succeeded on Apollo only due to NASA's constant close direction, without which we would have gone off on unpredictable tangents. For shuttle, NASA needed a contractor that would play ball, keeping the program alive in a hostile Congress, and it did not trust Grumman management, without Lew Evans, to do that.

North American Aviation's comeback recovery from the Apollo fire had built personal rapport with NASA management as both the agency and contractor worked closely together to save the program. In that sense a tragic failure sowed the seeds of North American's later success. I had long observed at meetings that NASA's Apollo management seemed more comfortable at North American than at Bethpage, showing closer sharing of viewpoints and preferences and never having to worry about being surprised by its contractor. I believed they considered us to be standoffish and erratic by comparison. For example, the chaotic LM-1 CARR would never have occurred at North American, which was far more aware of NASA's preferences in meeting accommodations. NAA's Apollo conference facilities in Downey remained far more comfortable and accommodating to NASA than Grumman's until well after LM-1's delivery. It was possible that our hubris and customer neglect were major factors in losing the space shuttle competition. NASA's decision left me disillusioned with the agency and with the government's procurement process. Past quality and performance seemed to count for little, even on a high-stakes, world-renowned program like Apollo. "What have you done for me lately?" was the only thing that counted.

North American Aviation held competitions for major space shuttle subcontracts, and Grumman won the subcontract to design and build the wing. This substantial job kept us closely involved with the shuttle program and maintained our engineering know-how on the latest spacecraft design and manufacturing techniques.

I had matured and broadened in outlook in the ten years from our winning LM proposal to our losing bid for shuttle. On LM I believed that if we had the best, most ingenious, and logical technical design and proposal we would win—simple as that. On shuttle I saw a far more complex array of factors, involving personalities, images, finances, management capability, and politics—technical performance was just one issue among many. My loss of innocence emphasized events beyond my control or influence as affecting winning or losing. It is a good thing I was not so worldly when it counted, on the LM proposal.

In April 1973, after useful discussions between Grumman president Jack Bierwirth and Secretary of the Navy John Warner, the F14 problem was resolved in a way that allowed Grumman to survive.[5] I became Grumman's director of Space Programs and took on the challenge of keeping Grumman in the space business. This was destined to be a long, difficult, and largely unrewarding struggle, although we succeeded in obtaining a low level of involvement with NASA and air force space programs. At first I fought to overcome feelings of cynicism and disappointment, but that soon faded and I adopted a more realistic attitude of building upon our existing talents and assets and forgetting the past except where it could contribute directly to our current ventures. The glory days of Apollo receded into Grumman's history, but I was involved in her present and future.

Epilogue

The Legacy of Apollo

What did we get from putting men on the Moon? Skeptics may say that all we got was a few hundred pounds of moon rocks and closeup looks at a lifeless discard pile, the target of eons of bombardment by random meteors. Plus a hard-fought victory for prestige in the cold war with the Soviet Union and an unforgettable picture of a lustrous blue-and-white Earth rising over the Moon's parched grayness. Yes, we got all of these, but I say we got much more—nothing less than a revolution in our world view. Apollo raised our vision and aspirations outward and upward, toward the far reaches of the universe and of human achievements.

The Apollo crucible also taught us things about ourselves, as a nation and a people. We saw the power of teamwork and focused goals applied to achievable peacetime ends. We developed advanced techniques to manage and coordinate huge efforts and to unleash the power of workers' productivity and ingenuity, which produced long-term improvement in America's global competitiveness. Bold enough to risk failure in full public view, the Apollo program reflected the confident, expansive mood of our free, prospering post–World War II society before the tragedy of Vietnam.

The legacy of Apollo has played a major role in raising America to leadership in a global economy. I saw this on a personal level and watched it diffuse through the general practice of management. Apollo showed the value of (1) quality in all endeavors; (2) meticulous attention to details; (3) rigorous, well-documented systems and procedures; and (4) the astonishing power of teamwork. I applied these precepts directly to Grumman's aircraft programs when I was vice president of Engineering. They have since become the main thrust of modern management practices, developing into widely used techniques, such as total quality management, computer-aided design and manufacturing, employee empowerment, and design and product teams, to name but a few.

The Apollo program set NASA on a course of sustained space exploration and exploitation during the last three decades of the twentieth century. The intensity and focus of Apollo helped establish the level and pace of NASA's space exploration activities. Its cumulative effect has changed how we see ourselves and our place in the universe.

Space technology has made possible instant worldwide communications and has spawned huge new industries using space-based or -derived technologies. Global communications and TV reception routinely take place via satellite, linking all peoples with an intimacy unimaginable to past generations. Cellular telephones have revolutionized the way the world does business and enabled undeveloped nations to leap-frog the installation of expensive ground-based infrastructure. A satellite-based Global Positioning System provides accurate knowledge of geographical position anywhere on Earth, greatly enhancing the safety of all modes of travel and the accessibility of every spot on the globe. These advances in space technology have resulted in the concept of the "global village," an enhanced recognition of mankind's interdependence and need for peace and brotherhood.

Observations from orbit have greatly expanded our knowledge of Earth and its environment. Our understanding of weather, climate, the oceans, the atmosphere, geology, nature, and endangered species has been multiplied and quantified by manned and unmanned measurements and observations from space. Many practical benefits have flowed from this information. Improved weather forecasting, for example, has saved lives and reduced property losses in natural disasters. Global climate modeling has helped us understand long-term natural trends and the extent to which human activities can affect them, thus making possible strategies to reduce problems such as atmospheric ozone depletion and global warming—problems not even recognizable prior to the availability of data from space. High-resolution imaging of the Earth at different wavelengths provides accurate information on land-use patterns, crop health and yields, erosion and fire damage. This valuable information is now being obtained and marketed commercially by private companies.

The geologic exploration of the Moon during Apollo was extensive, considering its pioneering aspect. Apollo excursions, samples, and surface and orbital measurements provided in-depth knowledge about the origins and composition of the Moon and its relation to Earth. The Moon was formed at the same time as Earth, about 4.5 billion years ago, soon after the formation of the Sun. Apollo provided a time line of major events in the Moon's geological history and showed both similarities and differences with Earth.

The Apollo missions highlighted human adaptability, demonstrating that people could work effectively in space under unique and hostile environmental conditions. The Skylab and Mir Earth orbital space stations further determined human capabilities and limits in space, as will the planned International Space Station (ISS). The ability to live and work in space for long periods of time is essential for manned missions to far places, such as Mars and beyond, and for colonization of the Moon or planets. A permanent human presence in Earth orbital space is also needed to fully exploit the space environment, using its high vacuum, low gravity, unobstructed sunlight, and unique perspective of Earth for whatever useful applications humans can devise. The ISS could be a precursor to manned planetary exploration—the first steps toward making the vision of *Star Trek* a reality.

NASA's extensive program of unmanned exploration of the solar system has yielded detailed knowledge about the Sun, the planets and their moons, and the solar and interplanetary environments. Pioneer, Explorer, Viking, Mars Pathfinder, and other programs have greatly increased the data on the solar system in only two decades. We can now compare and contrast Earth with every other planet and understand the major factors making the differences. The closeup views of far away worlds such as Mars, Jupiter, and Saturn, and their infinitely varied moons, has further stimulated our imaginations and desire that someday mankind shall visit these forbidding, exotic places.

Astronomical observatories and instruments in space have widened our window to the universe and caused ferment in cosmology and the theories of creation. The orbiting astronomical observatory, Grumman's pioneering large telescope in space, was succeeded by the Hubble Space Telescope, which is providing penetrating views to the farthest reaches of the universe. Other advanced observatories in orbit are probing other regions of the electromagnetic spectrum to astronomical distances: including the Shuttle Infrared Telescope Facility, the Compton Cosmic Ray Observatory, and the Chandra X-ray Observatory. At the long-wavelength end of the spectrum, very large ground-based radio telescopes, such as the facility in Areceibo, Puerto Rico, augment the data gained from space observations and keep watch for intelligent signals from remote galaxies and stars. The unprecedented flood of data has challenged cosmologists as never before, resulting in fascinating visions of the origins, extent and future course of stars, galaxies and the entire universe. It has stimulated the quest for a unified theory to explain the origin of the universe that can be verified by the observations.

The cumulative effect of these advances in human knowledge and technology resulting from the space program has been revolutionary and pervades humanity's contemporary view of ourselves and our world. We are far more accurately informed about our Earth and its place in the solar system and the universe than any previous generation. The once-exalted view of humankind as central to the universe has shrunk to a more humble but treasured one. Evidence has mounted that other planets are orbiting other Suns and that life on Earth is not unique in the universe. Statistics suggest that many millions of living worlds exist throughout creation, inhabited by a dazzling variety of life-forms and intelligent creatures, whose characteristics are a fertile field for speculation and imagination. What a grand universe the exploration of space has already opened up for us, in such a short span of time!

To have contributed a measure to this increase in cosmic knowledge and understanding is immensely satisfying and rewarding. It was my privilege to play a role in the greatest engineering adventure of the twentieth century. For that opportunity, and for the many friends and colleagues who helped our great enterprise succeed, I remain profoundly grateful. I wish my children and grandchildren and their generations the joy of continuing this exciting quest. As they look up at the Moon's glowing, seductive face, I hope they'll be inspired to still greater efforts by what we have done. Perhaps technology will let them look down on the Moon through their own computers, zooming in on each of the six lunar modules that sit in timeless isolation astride the foot-printed dust of a once-active lunar base. Why, I can see them up there with my naked eyes—can't you?

Notes

Chapter 1. A Difficult Delivery

1. The categories were Not Valid Chit (mistaken), Explanation Satisfactory, Documentation Correction Required, Retest or Replacement Required, and Unresolved.

2. Thomas J. Kelly, LM Meeting Notebook, bk. 6, 21 June 1967, 150. LM Meeting Notebooks are in the possession of the author.

3. Ibid., 152.

Chapter 2. We Could Go to the Moon

1. Richard Thruelsen, *The Grumman Story* (New York: Praeger, 1976), 286; Loyd S. Swenson Jr., James M. Grimwood, and Charles C. Alexander, *This New Ocean,* NASA History Series, NASA SP-4201 (Washington, D.C.: GPO, 1966), 137.

2. Courtney G. Brooks, James M. Grimwood, Loyd S. Swenson Jr., *Chariots for Apollo,* NASA History Series (Washington, D.C.: GPO, 1979), 26–29.

3. Charles Murray and Catherine B. Cox, *Apollo: The Race to the Moon* (New York: Touchstone Books, Simon & Schuster, 1989), 75–83.

4. The name "lunar excursion module" (LEM) was used by NASA until 1967. Because their Public Affairs Office thought that "Excursion" had a frivolous connotation, they shortened the name to "lunar module" (LM), which was still pronounced "lem." For simplicity and consistency, I refer to the manned lunar landing spacecraft as the lunar module (LM) throughout this book.

5. The original and most persistent (and effective) proponent of LOR was John Houboult, whom I met for the first time at this meeting. The usually visionary Faget at first opposed it, but he later became convinced of its merits. By the time of our meeting, he and Gilruth were fully converted LOR proponents.

Chapter 3. The LM Proposal

1. Grumman Aircraft Engineering Corporation, *Project Apollo—Lunar Excursion Module Proposal* (Bethpage, N.Y.: Grumman Aircraft Engineering, 4 Sept. 1962), 1–51. Available at Grumman History Center, Bethpage, N.Y.

2. Ibid., 1–52.

Chapter 4. The Fat Lady Sings

1. In 1961 Grumman invested $10 million in Plant 25, the new Engineering building, to provide office space if either the F111 subcontract or the LM was won. (Both were.)

2. A fine restaurant just outside Grumman's southern boundary fence.

3. Joseph G. Gavin Jr. to author, May 1997.

4. Brooks et al., *Chariots for Apollo,* 111–14.

5. In September 1967, North American Aviation was acquired by the Rockwell Manufacturing Company, and its name was changed to North American Rockwell Corporation. Later this was changed to its present corporate name, Rockwell International. In late 1996 Rockwell International sold its aerospace business to the Boeing Company, which now operates the facilities in Downey and Seal Beach, California, where the Apollo program activities took place.

6. Brooks et al., *Chariots for Apollo,* 41–42.

Chapter 5. Engineering a Miracle

1. The four roommates later bought the house, and as each got married, Rathke bought his share until he was sole owner. His bride Winifred, whom he met at Grumman, moved in when they were married.

2. The TFX cancellation also made Bill Rathke, who was Grumman's project engineer for the Missileer airplane, available for the LM program.

3. The command module needed prone position contoured couches to enable the crew to withstand four Gs during liftoff and eight Gs during atmospheric reentry.

4. Brooks et al., *Chariots for Apollo,* 137.

5. Michael Collins, *Carrying the Fire,* 2d ed. (New York: Farrar, Straus and Giroux, 1989), 339.

6. Gavin said that more than fourteen thousand LM test failures were recorded over the life of the program, of which only twenty-two were still unexplained at program's end. Joseph G. Gavin Jr., interview by author, Washington, D.C., May 1998.

7. Kelly, LM Meeting Notebook, bk. 1, 17 January 1964, 27.

8. Brooks et al., *Chariots for Apollo,* 159.

9. North American Aviation had been working under a loosely defined letter contract since late 1961. In August 1963 seven months of intensive negotiation concluded with the approval of a definitive contract for $934.4 million. Brooks et al., *Chariots for Apollo,* 132.

10. Ibid., 56.

11. Ibid., 136.

12. Kelly, LM Meeting Notebook, bk. 1, 16 January 1964, 21–26.

13. For example, the level 2 diagram for the ascent propulsion system showed the tanks, rocket engine, valves, regulators, plumbing, and electrical control inputs and outputs.

Chapter 6. Mockups

1. Charles R. Pellegrino and Joshua Stoff, *Chariots for Apollo: The Making of the Lunar Module,* 1st ed. (New York: Atheneum, 1985), 57.

2. Brooks et al., *Chariots for Apollo,* 162.

Chapter 7. Pushing Out the Drawings

1. Murray and Cox, *Apollo,* 61–65.

2. Kelly, LM Meeting Notebook, bk. 1, 13 May 1964, 122–23.

3. Ibid., bk. 2, 22 September 1964, 43–44.

4. Thruelsen, *Grumman Story,* 241.

5. Kelly, LM Meeting Notebook, bk. 4, 18 April 1966, 127.

Chapter 8. Trimming Pounds and Ounces

1. Johnson's list included some pretty "far-out" items, such as cutting the LM crew down to one man and eliminating system redundancy—for example, making LM passive during lunar-orbit rendezvous by deleting the rendezvous radar. These were probably not meant as serious contenders but served to impress Grumman with how seriously NASA took LM's weight problem and to stimulate us to question all the "givens" of the design and the mission plan.

2. Kelly, LM Meeting Notebook, bk. 2, 9 October 1964, 56–57; 9 November 1964, 77–80.

3. Ibid., 16 March 1965, 152.

4. Ibid., bk. 3, 8 April 1965, 18–19.

5. The engineer "cognizant of" (i.e., in charge of) technical performance of a subcontracted or purchased item.

6. Kelly, LM Meeting Notebook, bk. 2, 26 February 1965, 139.

7. Brooks et al., *Chariots for Apollo,* 173–74.

8. Range rate could be calculated from successive range measurements.

Chapter 9. Problems, Problems!

1. Equivalent to one sugar cube–sized volume of helium per day at sea-level temperature and pressure.

2. The "Grumman ironworks" referred to the frequent ability of Grumman airplanes to survive combat and return safely to base even though shot full of holes in battle. Grumman's rugged planes were considered worthy of an ironworks. "Sterling on silver" refers to a cherished quotation from a navy admiral that "the name Grumman on an airplane is like sterling on silver."

3. Brooks et al., *Chariots for Apollo,* 256–60.

4. Murray and Cox, *Apollo,* 146–51, 179–80.

5. Specific impulse equals thrust per propellant flow rate; a measure of rocket engine efficiency, roughly equivalent to miles per gallon for automobile engines.

6. Brooks et al., *Chariots for Apollo,* 245–46.

7. Ibid.

8. Ibid., 171–72, 211.

9. Apollo 18 was canceled and never flown. The program ended with Apollo 17.

10. Kelly, LM Meeting Notebook, bk. 8, 5 October 1968, 55; 8 October 1968, 57; 12 October–8 November 1968, 59–69; 21 March 1969, 93–94.

Chapter 10. Schedule and Cost Pressures

1. Years later, NASA Administrator James Webb revealed that he had arrived at this amount by tripling the $8 billion estimate given to him by the Apollo program management and other NASA experts. He was sure no one knew how to price a program of manned lunar exploration. (James Webb, after-dinner speech, Washington, D.C., c. 1972).

2. Brooks et al., *Chariots for Apollo,* 167, 177, 189.

3. Thruelsen, *Grumman Story,* 323–24.

4. Jack Buxton told me that he and Gavin were summoned to a meeting after Evans heard the review team's summary briefing. They were told to bring the latest LM program organization chart. Evans spread the chart on the desk before him, and starting with the top box, "J. Gavin–VP, LM Program Director," he checked off those to be retained and crossed out those to be eliminated, adding new position boxes and changing reporting lines as he went.

5. The critical path was that path through the program evaluation and review technique schedule network that showed the greatest schedule slippage. It was often called "the long pole in the tent."

6. Brooks et al., *Chariots for Apollo,* 201.

Chapter 11. Tragedy Strikes Apollo

1. Brooks et al., *Chariots for Apollo,* 114–18.

2. Astronauts Elliot See, Charles Bassett, and Theodore Freeman were killed in T-38 crashes, the former two as they approached the McDonnell Douglas plant at the St. Louis Airport in bad weather, on their way to attend a Gemini Project meeting.

3. The mission had been designated Apollo-Saturn (AS)-204, but a revised designation system was in the works to number the Apollo missions sequentially, with AS-204 becoming Apollo 1. After the fire, this nomenclature was adopted by NASA. Due to program changes, there were no Apollo 2 and 3 missions; the next mission flown was Apollo 4.

4. Absolute pressure is referenced to zero, whereas differential pressure is referenced to the local atmospheric pressure, which is 14.7 psia at standard sea-level conditions. Thus a tank rated at 2000 psi differential pressure could contain oxygen at 2000 psia in space (zero atmospheric pressure), or 2014.7 psia at sea level.

5. At the crew's suggestion, Shea had planned to witness the "plugs out" simulated launch test inside the command module, curled up at the foot of the astronauts' couches. He dropped this idea when the test team was unable, on short notice, to rig up a temporary fourth communication channel and headset for his use.

6. Murray and Cox, *Apollo,* 209–12, 215–20.

7. Brig. Gen. Carroll H. "Rip" Bolender of the U.S. Air Force became ASPO's lunar module manager early in 1968. Kenneth S. "Ken" Kleinknecht, formerly NASA's Mercury Project manager, was his ASPO counterpart as command and service module manager.

8. Brooks et al., *Chariots for Apollo,* 224.

Chapter 12. Building What I Designed

1. Kelly, LM Meeting Notebook, bk. 6, 17 February 1967, 82.

2. Apollo 18 was ultimately canceled, leaving Apollo 17 the last lunar landing mission in December 1972.

3. Kelly, LM Meeting Notebook, bk. 7, 19 February 1968, 6/67–3/68, 132.

Chapter 13. First LM in Space: Apollo 5

1. This was the standard mission command hand-off rule for all the Apollo flights.

2. Brooks et al., *Chariots for Apollo,* 241–44.

Chapter 14. The Dress Rehearsals: Apollos 9 and 10

1. It was, as Pete Conrad told me, the difference between commitment and involvement. With regard to a bacon-and-egg breakfast, the pig is committed, whereas the chicken is involved.

2. Brooks et al., *Chariots for Apollo,* 290–91.

3. Ibid., 256–60.

4. The Moon was found to be nonhomogeneous, with several internal mass concentrations ("mascons") that resulted in a very complicated gravitational field close to the surface.

5. Brooks et al., *Chariots for Apollo,* 244–46, 286–87.

6. CapCom and the astronauts had a secure, guarded channel to which they switched whenever they needed privacy, which always included cases of crew sickness or other problems.

7. Edgar M. Cortright, ed., *Apollo Expeditions to the Moon,* NASA SP-350 (Washington, D.C.: GPO, 1975), 190.

8. In one of the more outlandish examples of space jargon, the backpack was called the portable life-support system, or PLSS, pronounced "pliss."

9. Brooks et al., *Chariots for Apollo,* 299

10. Ibid., 299–300.

11. Ibid., 303–12.

Chapter 15. One Giant Leap for Mankind: Apollo 11

1. Pellegrino and Stoff, *Chariots for Apollo,* 168–69.

2. The "a" was not audible in the transmission to Earth, although Armstrong claimed he spoke it. Andrew Chaikin, *A Man on the Moon: The Voyages of the Apollo Astronauts* (New York: Penguin Books, 1994), 209.

3. This description of the "feel" of ascent was given to me by astronauts Dave Scott and Jim Irwin, in conversation following their Apollo 15 mission, Houston, Texas, c. September 1971.

4. Brooks et al., *Chariots for Apollo,* 353.

5. Ibid., 340.

6. Collins, *Carrying the Fire,* 412.

Chapter 16. Great Balls of Fire! Apollo 12

1. The command module had switched to battery power, which was a small source intended only for reentry.

2. Chaikin, *Man on the Moon,* 235–39.

3. After Apollo 11's successful mission, astronaut Jim McDivitt became Apollo spacecraft program director, succeeding George Low, who moved up to assistant director of the Manned Spacecraft Center in Houston.

4. Murray and Cox, *Apollo,* 372–82.

5. Doppler shift is the apparent shift in frequency of light, sound, or radio waves emanating from a moving object as viewed by a stationary observer. Thus the whistle of a moving train sounds progressively higher pitched as it approaches and lower pitched as it recedes.

6. This confirmed the design requirement that we and NASA had included from the outset that the lunar module must be capable of landing from one hundred feet under instrument flight rules in anticipation of a lunar-dust visibility problem.

7. Chaikin, *Man on the Moon,* 250–60.

8. Ibid. 279–80.

Chapter 17. Rescue in Space: Apollo 13

1. This was a significant result of the Apollo Mission Planning Task Force (AMPTF) study led by Grumman for NASA in early 1964.

2. Jim Lovell and Jeffrey Kluger, *Lost Moon: The Perilous Voyage of Apollo 13* (New York: Houghton Mifflin, 1994), 250–57. Reissued in paperback as *Apollo 13* by Pocket Books division of Simon & Schuster, 1995.

3. Lovell and Kluger, *Lost Moon,* 282–85.

Chapter 18. The Undaunted Warrior Triumphs: Apollo 14

1. Chaikin, *Man on the Moon,* 337–41.

2. Slayton's medical problem was intermittent heart fibrillation.

3. Under Slayton's unofficial system, a backup crew became the prime crew two missions later.

4. In this endeavor, the astronauts were following a rich tradition of previous great explorers. I am struck by the similarities of the astronauts' lunar geology studies with the private tutoring as a field naturalist, biologist, botanist, and navigator that Meriwether Lewis obtained in 1803 before embarking upon his expedition to the northwest Louisiana Territory. (See chapter 7 of Stephen Ambrose's *Undaunted Courage* [New York: Simon & Schuster, 1996].)

5. Chaikin, *Man on the Moon,* 347–52.

6. Ibid., 352–54.

7. Solder balls inside the case were a not infrequent problem with hermetically sealed switches and instruments. A small tube in the case was used to evacuate the air and refill with inert nitrogen; this tube was then pinched off and sealed with solder. If the pressure inside the case during sealing was negative, solder could be sucked inside. It would usually be detected at the factory by vibrating or shaking the switch.

8. Chaikin, *Man on the Moon,* 357–60.

9. Ibid., 374–75.

10. Thomas J. Kelly, annotations on personal copy of Apollo 14 Flight Plan, February 1971. In possession of author.

Chapter 19. Great Explorations: Apollos 15, 16, and 17

1. Chaikin, *Man on the Moon,* 412–15.

2. The Moon as we see it consists primarily of the dark maria, known from earlier missions to be created by vast lava flows about 3.85 million years ago, and the bright highlands, origins unknown. Both areas are pocked with impact craters of widely varying sizes, and littered with impact ejecta (rocks and boulders). See also Chaikin, *Man on the Moon,* 452–56.

3. The gimbal control system controlled electric motor–driven actuators which pointed the large SPS rocket engine. For stable flight when thrusting, the engine was pointed at the spacecraft's center of gravity.

4. Chaikin, *Man on the Moon,* 456–62.

5. Ibid., 463–75.

6. Skurla gave the guilty parties a few days off until Petrone's wrath subsided.

7. Scientists dated the glass beads as 3.5 billion years old, not far removed in geologic time from the extensive lava flows 3.85 billion years ago that created the vast lunar maria.

8. Scientists studied these boulders for years, and from them derived a reconstruction of the violent events that occurred when the huge Serenetatis asteroid slammed into the Moon 3.8 billion years ago, thrusting up the Taurus Mountains and the Massifs and showering debris thousands of miles across the surface.

9. The message on the plaque was "Here man completed his first explorations of the Moon, December 1972 AD. May the spirit of peace in which we came be reflected in the lives of all mankind."

10. Chaikin, *Man on the Moon*, 516–30, 535–45.

Chapter 20. Our Future Slips Away

1. This capability was known as "cross-range" because it permitted moving the reentry trajectory in the direction perpendicular to (across) the downrange direction of the spacecraft's original orbit.

2. The phase B design studies would be followed by phase C, in which NASA would select and define a design approach for the spacecraft competition, followed by the industry competition (phase D).

3. Thruelsen, *Grumman Story*, 362–71.

4. In NASA's source selection process, the SEB does not recommend a winner but evaluates the relative strengths and weaknesses of the proposals and rank orders them using a weighted scoring system for the technical and management proposals. It also evaluates the relative validity and realism of the cost proposals. The SSA considers the SEB's report as a major input to his deliberations but makes the selection based upon the overall best interests of the government.

5. Thruelsen, *Grumman Story*, 372.

Index

Page numbers in boldface refer to illustrations.